Kyle Cathie Limited

Nick Sandler and Johnny Acton

photography by Peter Cassidy

PRESERVED

FOREWORD

I've long been a fan of Johnny and Nick's culinary double act, not least because I know they share my passion for 'real' food – ideally locally sourced and in season. It's great to see them combine here on such a worthwhile subject, and make the various alchemies of the preservers' art accessible and enticing.

As a stickler for cooking with rhythm of the seasons, it may seem paradoxical to advocate processes that allow you, for example, to eat raspberries in December. But in many ways, I see preserving as a celebration, not a denial, of seasonality. I'd rather eat my own raspberry jam (or Johnny's and Nick's mango chutney – see page 170) on Christmas Day, than buy fruit and vegetables flown in from half way round the world.

For those of us who grow their own vegetables and fruit, and rear our own livestock, home preserving is an essential tool in the box of good smallholder practice. It's a way of making the most of the glut of the harvest – bottling it, either literally or metaphorically, for later consumption. But even if you don't have the time and space for raising your own food, you can still tap into the fine fresh produce of those who do (ideally by buying direct from producers at farmer's markets) and dabble successfully in the art of preserving. The fruits of such labour, apart from being delicious, add immeasurably to the home cook's sense of self-sufficiency – and that's a feeling you can't put a price on.

Preserving is about intelligent home economy. It is a matter of maximising resources, using know-how both ancient and modern to get the most out of your food. But this doesn't imply parsimony or tight-fistedness. If understood and practised well, preserving techniques will never produce food that seems meagre, mean or bland. Instead they introduce a whole new range of flavours and textures to the cook's repertoire, which are as pleasing and exciting as they are diverse.

This is because the key ingredients and techniques for the principal preserving methods – salt and sugar, vinegar and alcohol, oil and fat, smoke and even plain air – all have radical transforming effects on the raw ingredients exposed to them.

With the judicious inclusion of herbs, spices and other aromatics, the range of tastes and effects that can be brought about by home preserving is dazzling – and often delightful.

The result is that techniques that began as fairly crude ways of preventing spoilage and laying down stores have developed into some of the higher culinary arts. And preserved foods that were once staples have become our favourite luxuries.

These days most of us rely on commercial processors for our preserved foods – whether it's jams or pickles, cured meats or smoked fish. Yet often the scale of these operations involve compromise on ingredients and corner cutting of technique – the inevitable consequences of industrialised food production. What Nick and Johnny have shown here is that the techniques involved are not difficult to master on a domestic scale – and that the process of learning them is both fun and rewarding. Most importantly, they show that the preserved foods you make at home, with the best ingredients you can lay your hands on, can and will be the finest you ever taste.

As an introduction to the subject, *Preserved* is first rate, full of recipes that are both achievable and delicious. Nick and Johnny evidently had a whale of a time exploring the world of food preserving – and there is every chance that, with their guidance, you will too.

Hugh Fearnley-Whittingstall

WHY PRESERVE?

It may sound grandiose, but in many ways the history of civilisation is the story of our progressive mastery of food preservation. After all, to be wild is to eat on the hoof. Learning to prolong shelf-life enabled our distant ancestors to migrate beyond regions where adequate fresh food was available all year round. Later, the construction of cities was premised on our ability to preserve grains, pulses and animal products. By freeing us from the previously universal imperative to hunt and gather, preserving permitted specialisation, giving birth to the arts and sciences. Later still, preserved foods enabled people to survive long sea journeys and thus to colonise the globe. One day, they may allow us to do the same in Space.

Anyone in any doubt about the centrality of food preservation to the way we live should pay a visit to the local supermarket. Almost all of the ingredients we use in our cooking have gone through some kind of preservation process. Take everything cured, smoked, dried, hung, fermented, candied, canned, tinned, frozen, pasteurised, irradiated or vacuum-packed from the average shopping basket, and you aren't left with much. Just fresh fruit and vegetables and a few fish and poultry products. And probably not even those, unless you go all-out organic.

So food preserving is important, both culturally and historically. Indeed, our distinctive national tastes almost always reflect the preserving techniques dominant in the relevant bit of the world (Scandinavians, for instance, often have a liking for the taste of ammonia as a consequence of the rather drastic methods used by their forebears to preserve fish). But the question remains, why bother to do it ourselves?

TAKING CONTROL OF YOUR FOOD

There are many reasons why home-preserving is a satisfying and worthwhile undertaking. Among the most topical is a growing awareness of the drawbacks of mass-produced food. Some of these are health related, others concern taste. In terms of the former, many people are understandably worried by the additives employed in large-scale food manufacture. When it comes to quality, the end products are frequently compromised by two entrenched habits of the food giants. The first is the use of short-cuts, as time is money in the business world. 'Smoked' meats, for example, are often simply infused with 'liquid smoke' rather than being produced according to patient, traditional methods. This inevitably has a deleterious effect on taste. The second problem with supermarket food is 'averaging' – in other words, catering to the average rather than the individual palate. You may like your kippers smokier or your pickled onions sharper than Mr and Mrs 2.4 kids, but if you buy commercial produce, you just have to lump their preferences. With home-preserving, you can tailor your food just the way you like it.

Beyond the problems inherent in mass-production, several trends point towards home-preserving. One is the gardening boom. Now that we've taken to growing all these cucumbers and tomatoes, what are we to do when they arrive in sudden gluts? The learning of a few simple preserving techniques opens up a world of possibilities. Then there is the vogue for sophisticated home-cooking. If the transformations that give your ingredients their distinctive character have already taken place by the time you get hold of them (as when you buy a side of smoked salmon), cooking with them will be far less fulfilling than if you prepare them from scratch. With home-preserving, there can be no such problem. The satisfaction goes right to the bottom.

A WORLD OF FLAVOUR

Another significant trend is an increasing willingness to experiment with new tastes. In recent years, eating has gone truly global. Home-preserving provides access to a vast range of 'foreign' delicacies, many of which are not available in the shops. *Preserved* celebrates this international diversity to the full.

Many other incentives could be added to the list. There is the sheer joy of bringing about amazing and delicious transformations (by concentrating flavours, for example). Then there are the economic benefits – superior preserved foods are much cheaper to make than to buy. They also tend to be aesthetically pleasing, and so make excellent gifts. Plus there is the atavistic thrill of mastering techniques that have been so important to human survival. But above all, preserving is fun.

THE HISTORY AND GEOGRAPHY OF FOOD PRESERVING

Most of the great preserving methods were stumbled upon by accident, too long ago to date with any accuracy. A desperate hunting party found a dead animal lying on a salt-pan, and found that its flesh tasted curiously fresh. A herdsman left a skin of milk in the sun by mistake, later discovering that it had curdled into something rather good. A farmer put some leathery old cabbages in a pot of sea water, forgot about them and returned to find them shrunken but delicious. But it took acute observation and the sharing of information through language to translate these discoveries into duplicable techniques. Our forebears may not have understood, in our terms at least, why subjecting their food to certain processes kept it wholesome. For them it was a kind of magic. But notice it they did.

CLIMATE AND LOCATION

The particular preserving techniques which took hold in a given part of the world depended on climate and the *modus vivendi* of the inhabitants. Among the nomadic peoples of the deserts and steppes, for whom growing vegetables was clearly a non-starter, doing clever things with milk was paramount to survival. In the Arctic and in mountainous areas, food froze automatically in winter unless strenuous efforts were made to prevent it. The cool, dry atmosphere of such zones also lent itself to air drying, giving the world such delights as bresaola. In the Mediterranean, Arabia and parts of the Americas, similar results could be obtained by leaving food in the fierce summer sun, although the products were different, typically consisting of fruits and chillies. Meanwhile, in temperate latitudes where conditions were too moist for effective drying, the accents were on smoking, salting and pickling. And in Japan, Korea and other mountainous parts of the Far East, huge populations were built on fermented pulses and fish, in defiance of the scarcity of cultivable land.

In northern climes, autumn was the cue for a frenzy of preserving activity for thousands of years. Few households owned enough land to provide winter fodder for their livestock, so non-breeding animals were invariably slaughtered before the big freeze. Time was of the essence, and all family members, together with the neighbours, would be enlisted to salt, smoke

and dry the flesh and convert the leftovers into sausages. After the harvest, vegetables, fruits and fungi all had to be dried or pickled rapidly to prevent them going off, so their preparation was also a communal undertaking.

In Europe, the pattern went unchanged for generations. The only exception was the arrival of sugar, first introduced in minute and prohibitively expensive quantities by crusaders returning from the Middle East. Its preserving qualities were quickly recognised, but despite the efforts of entrepreneurial farmers in the Canary Islands and elsewhere, it remained beyond most people's budgets until the transformation of the West Indies into one giant sugar plantation during the seventeenth and eighteenth centuries. Suddenly, a whole new range of preserving possibilities opened up to ordinary folk, from jams and marmalades to candied fruits and sugared nuts.

A MORE SCIENTIFIC APPROACH

The greatest revolution in preserving, however, occurred in 1861. In that year, Louis Pasteur published a paper which finally solved the age-old mystery of why food was inclined to go bad. The air, he explained, was teeming with organisms invisible to the naked eye. He had proved this by drawing everyday air through clean guncotton filters and then examining them with a microscope. He had also dropped one of these filters into a sterilised jar of nutritious 'soup' and watched the contents start to putrefy. At last people knew the real enemy. Food preserving had previously been an uncomfortably hit-and-miss affair, and the misses had frequently had lethal consequences. But now, with a proper understanding of hygiene and the advent of scientifically sound methods of bottling and canning, safely preserved food became available to the masses. This applied not only to industrially processed products, but also to those made in the home.

We, in our shrunken modern world, are the lucky inheritors of all these techniques. We don't have to live in a desert to eat dried dates or up an oriental mountain to enjoy a steaming bowl of miso soup. But an appreciation of the rich history behind the many preserved foods that can be successfully made at home can only make them taste better.

DRYING

Where would the Italians be without pasta, or the Chinese without noodles? In serious trouble is the answer. The same could be said of the Egyptians without dried lentils, or rather less kindly, of the residents of Fulham or San Francisco without sun-dried tomatoes. Many dried foods are so familiar that it's easy to forget that's what they are.

Prehistoric humans inevitably came across wind-fallen fruits that had dried in the sun. They would have noticed that they didn't look too bad and tasted sweeter and chewier than average, and then that they remained edible for much longer than their fresh equivalents. Any number of forgetful moments around the home would have taught them similar lessons regarding meat.

The systematic removal of moisture from food is the oldest and simplest preserving method of all. It works because potentially contaminating organisms are sunk without water. It also concentrates flavour, sometimes to an extreme degree, as in the pungent delicacy euphemistically known as Bombay Duck. Made on the west coast of India by drying small fish in the hot sun, Bombay Duck is intense, salty and almost mineral in texture, but it keeps forever. Herodotus, writing in the fifth century BCE, described the Egyptians processing small fish in the same manner. Alexander the Great's troops noticed the residents of Baluchistan in present-day Pakistan making a version of Bombay Duck into flour and even feeding it to their animals.

Their neighbours in Persia and Afghanistan have been adding dried apricots, dates and mulberries to their stews since time immemorial. Central and South Americans have been air drying strips of meat for just as long. They called the end product *charqui*, which gives us the modern word 'jerky'. Salt cod was so important historically that the slave trade would arguably have been impossible without it. It was sustained by a triangle in which huge quantities of salted fish were purchased in Newfoundland and New England with commodities from the West Indies like rum and molasses. This cod was then used to buy slaves in West Africa, and to keep them alive once they had arrived in the Caribbean. The sugar plantation owners paid for them with the very products that were in such demand on the North Atlantic seaboard. Thus was the triangle completed, and the traders repeated it indefinitely, taking a substantial profit at every turn. Dried pulses were the chief source of protein in Europe and elsewhere for many centuries, as commemorated in the nursery rhyme: 'pease pudding hot, pease pudding cold, pease pudding in the pot nine days old'. And dehydrated foods, being light and almost imperishable, have long been vital to armies.

Many of the best dried foods once depended on desiccating mountain breezes or fierce tropical sun, conditions scarcely prevalent in damp corners of the globe like Britain. But with a bit of ingenuity you can now compensate for adverse local conditions. A drying box will prove invaluable. Instructions for building one appear in the recipe for biltong on page 14.

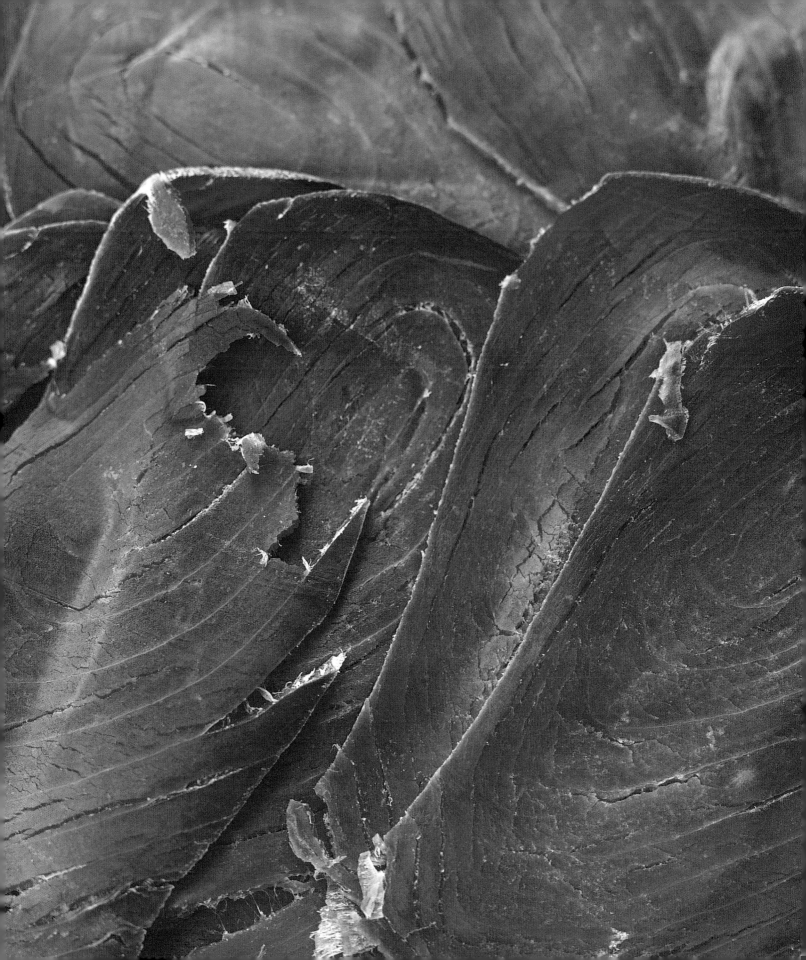

BILTONG

Mention biltong to *émigré* Southern Africans and their eyes will start to water with nostalgia. Dark, chewy and frankly pretty tough, this air-dried spiced meat is an acquired taste, but once acquired it is never forgotten. Americans already have a head start through their predilection for beef jerky, but never make the mistake of comparing the two in the presence of a South African!

The word 'biltong' is of Dutch derivation, 'bil' meaning buttocks and 'tong' meaning strip, but the Boers learned the technique through contact with the indigenous Bantu. They found biltong an invaluable source of protein during their long wagon trek across the African subcontinent, which began in 1836. The standard spicing is a dramatic blend of vinegar, pepper, salt, sugar and crushed coriander seeds. These commodities were readily available during the nineteenth century due to the combined efforts of wine-making Huguenot settlers and seafarers stopping off at the colony *en route* from the spice markets of the East.

Biltong can be made from several kinds of lean red meat. Kudu, impala and ostrich varieties are all popular in its homeland, but the standard form is beef. The difficulty that needs to be overcome is the duplication of the dry atmospheric conditions characteristic of southern Africa. This can be achieved with a measure of success by hanging the strips of meat on a line in a cool, dry place and placing a whirring fan nearby, but if the air is humid the biltong may still spoil.

A better alternative is to build a drying box. This is remarkably simple (see the instructions to the left), and you can use the box to dry other products – for instance, fruit.

MAKING A DRYING BOX
Use a large wooden or cardboard box (ours is 75cm/2½ft high by 45cm/1½ft deep by 60cm/2ft wide), completely sealed but for a few holes made in the sides towards the bottom and the top. Place a lit 60-watt light bulb inside, protected from dripping juices by a perforated piece of wood or cardboard suspended above. Keep the box out of the sun.

PREPARING THE BEEF

Get hold of some silverside of beef (or London broil as it is called in the US) and cut it with the grain into strips approximately 1cm (½in) thick and 15cm (6in) long. Cut away any excess fat, as it may otherwise turn rancid. Sprinkle the beef strips liberally with rock salt on each side and leave them for about an hour. Then scrape off the excess salt with a knife (do NOT use water). Place some cider or wine vinegar in a bowl, and submerge each piece of meat for a second or so before holding it up so the excess drips off. Then sprinkle all over with ground pepper and coriander seed.

THE DRYING PROCESS

The meat is now ready to dry. Suspend the strips towards the top of your box via meathooks or string. Switch on the light, and leave for three to four days, by which time the process should be complete. The bulb will produce warm, dry, rising air and the ventilation holes will help it to circulate.

Variations on the basic theme can be made by painting the meat strips with barbecue, Worcestershire, Tabasco or soy sauce between the vinegar and pepper/coriander stages.

SERVING AND STORAGE

Biltong is traditionally thinly sliced – in southern Africa they have special machines for the task. Store lightly wrapped in a cool, airy place or refrigerate. Eat within 6 months.

JERKY

From the American perspective, biltong is the South African version of jerky. For South Africans, jerky is American biltong. The two products do have a lot in common. Both were historically important in the diets of the relevant indigenous populations. They are similarly nutritious and addictive, and ersatz versions of both are increasingly sold and packaged commercially. But they are different enough to make it well worth your while experimenting to decide where you stand in the intercontinental dried meat debate. Jerky is typically made from thinner strips of meat than biltong. Store it in the same way, and eat within six months.

As mentioned in the introduction to this chapter, jerky is a corruption of the word *charqui*. There is some debate about the origin of the term, but the basic technique is continent-wide. The Quechua Incas of Peru cured alpaca meat in salt from local *salares* (salt flats), dried it in the altiplano desert sun and called it *ch'arki*. In Chile, the guanaco, another relative of the llama, was treated in a similar manner. Chilean miners were among the first on the scene during the California gold rush of 1849. There they took to making jerky from the local cattle and passed on the habit to their fellow prospectors. Meanwhile, the Indians of the North American plains had been drying strips of buffalo meat since ancient times.

TO MAKE JERKY

As with biltong, jerky can be made from almost any lean flesh, including fish. Turkey jerky has a particularly nice ring about it. But the instructions below are for beef jerky, the most popular variety of all.

> 1kg (2¼) trimmed lean beef (top rump is ideal)
> 250ml (9fl oz) soy sauce
> 2 tablespoons sugar
> 3 cloves of garlic
> ½ medium onion, finely chopped
> 1 teaspoon Tabasco sauce
> 2 tablespoons malt vinegar
> 1 tablespoon Worcestershire sauce
> A squeeze of lemon juice

First cut the meat, going with the grain, into strips about 5mm (¼in) thick.

Next, blend the marinade ingredients until smooth, add the meat and marinate for 6 hours minimum in the fridge.

Now you have a choice of drying methods:

SUN/AIR DRYING

Only try this if you live somewhere hot where the air is genuinely arid. Even then, wait for a breezy day. The easiest way to dry your jerky is to hang the strips on a clothes line, protected by some suspended muslin. It is ready when the meat is breakable but not yet brittle; this may take 3 or 4 days. But don't use this method for poultry – the risk of spoilage is just too high.

OVEN DRYING

Cover the floor of your oven with tin foil to catch the juices.

Preheat the oven to 80–90°C/180–200°F/gas mark ¼.

Lay the strips of meat on a wire rack and wipe off the drips. Then 'cook' for about 4 hours with the oven door slightly ajar, turning the meat over half way though. It may take slightly longer to reach the desired consistency – test it periodically.

SMOKING

If you want to try smoking your jerky, first have a look at the relevant chapter of this book. Then hot-smoke for at least 2 hours (testing it after this period) at about 90–95°C/200°F, or cold-smoke for rather longer (6–8 hours at 60°C/140°F), finishing it off in the hot-smoker or oven if necessary.

BRESAOLA

Bresaola is soft-salted and air-dried beef eaten raw. The original and best examples hail from the Valtellina mountains on the borders of Italy and Switzerland. Cut into thin, succulent, almost translucent, ruby-red slices and served with olive oil, lemon juice and parmesan, bresaola is one of the classic Italian starters.

TO MAKE BRESAOLA

- 1 large lean top rump (around 2kg/4½lb) tied tight with string
- 500ml (18fl oz) red wine (e.g. Chianti)
- 2 teaspoons ground red chilli powder
- 4 cloves of garlic, crushed
- 6 bay leaves, shredded
- 750g (1lb 10oz) coarse salt
- 1 tablespoon coarsely ground black pepper
- 10 sprigs of rosemary, roughly chopped
- 10 sprigs of thyme, roughly chopped
- 3 tablespoons sugar
- Enough muslin to wrap the beef
- Red wine vinegar, for washing

Place the beef in a large tupperware container and cover it with all the ingredients bar the muslin and vinegar. Massage them well in. Leave the meat to marinate in the fridge for 1 week, turning it over every day to ensure an even distribution of marinade.

After this period, brush the marinade off the beef and wrap it in muslin. Hang it in a dry, cool place for 1 month. It will drip for a day or two, so take appropriate measures to protect your floor.

The bresaola has matured when it feels firm to the touch. Once it is ready, wash it down with red wine vinegar, then dry it with a cloth. Store in the fridge, preferably in a container, for up to 1 month.

DRIED HERBS

By autumn, tarragon threatens to take over Nick's small urban garden. The Greek basil is bushy and pungent and mint is marching up the walls. The time has come for a drying orgy.

Dried herbs pack more power per teaspoon than fresh ones. Or at least they do until they become superannuated, provided you've prepared and packaged them correctly. You will have discovered this if you've ever tried making a familiar dish with the form you don't usually use. This is simply a matter of moisture loss: once herbs are dried, the oily constituents that give them their flavour are more concentrated. This has obvious storage benefits and allows you to enjoy your favourite varieties all year round. If you need further incentive, have a look at the cost of dried herbs in your local supermarket.

Herbs can be dried in the sun, on the stalk, in an oven or even in the microwave. Their versatility extends to sauces, marinades, seasonings and rubs. And there are classic mixes like bouquets garnis and *herbes de Provence* to enjoy. The latter can just as easily become *herbes de Rochdale* or *herbes de Cleveland, Ohio*.

BAG DRYING

This method is suitable for whole plants or large bunches of herbs. Only wash the herbs if absolutely necessary, draining them thoroughly on kitchen paper. Place them in a paper bag with their heads down and the base of their stalks protruding. Then tie some string around the neck of the bag and hang it in a suitably warm and airy place. Gravity will ensure that the flavoursome oils flow down into the leaves. When the leaves are dry (guide time 3 to 5 days), rub them off the stalks and store them in an airtight container. They should keep for a good year.

RACK DRYING

Take the leaves off the stalks and dry them on a fine-meshed rack in a dark room until brittle. Store like bag-dried herbs.

MICROWAVE DRYING

Place a dinner-plate-sized single layer of herbs in the microwave on a absorbent kitchen paper, then switch to 'defrost' for around 5 minutes. If they are not totally dry, give them another minute or two. Then store in an airtight container for up to 1 year.

HERBES DE PROVENCE

This is probably the most famous herb mix in the world. Flake together dried thyme, marjoram, oregano, rosemary, savoury, basil and tarragon to make a fantastic seasoning. Nick's celebrated roast chicken seasoning consists of 2 tablespoons *herbes de Provence*, 1 tablespoon paprika, 1 tablespoon sea salt, 1 tablespoon cracked black pepper, 1 tablespoon mustard powder and 2 teaspoons garlic powder shaken together and stored in a jar. Plaster this onto your bird, baste with olive oil and lemon juice, and you will be transported to the south of France.

BOUQUETS GARNIS

Take a twig of leafy bay, 4 stalks of thyme, 2 lengths of rosemary and some parsley, and tie them together tightly with some string. Make a few such bouquets and dry them using the paper-bag method (see left). They make great presents.

SMOKED ROSEMARY

This is well worth trying to give a novel twist to your roast lamb and/or potatoes. Simply smoke some rosemary in a cold-smoker for an hour, then dry it using one of the methods above.

DRIED MUSHROOMS

Forgive the self-promotion, but wild mushrooms are another big part of our lives (see our book *Mushroom*, also published by Kyle Cathie Ltd). The mushroom season in the UK is a fickle thing, so when the atmospheric conditions are right, which for most species means after late summer or autumn rain but before the first frosts, one must be ready to pounce. (Morels are a notable exception, appearing in the spring.) To add to the sense of urgency, many fungi deteriorate rapidly after harvesting unless something is done to prevent it.

Fortunately, the majority of species (puffballs, truffles and inkcaps are notable exceptions) dry extremely well. Some, like porcini, shiitake and morels, develop welcome new flavours and textures through the process. Dried mushrooms also have a beguiling sculptural quality, and a few jars of them give a kitchen plenty of atmosphere. And preparing them has a sound royal pedigree – Louis XIII of France was stringing morels on his death-bed.

All mushrooms should be brushed thoroughly ahead of drying to clear them of dirt and insects. The latter can secrete themselves in the most unlikely places. You may need to cut hollow-stalked varieties like morels in half to guarantee an earwig-free experience.

DRYING ON THREADS
This method is ideal for morels (the delectable pitted ones that look like mini brains and appear in late spring) and chanterelles (the apricot-coloured ones with gills that run half way down their stalks). Simply take a needle and a metre/yard of strong cotton and thread the mushrooms down its length, making sure there is a small gap between each one or they may meld. Hang them in a warm, well-ventilated area such as an airing cupboard, or over your Aga if you have one, and expect them to be dried after 2 or 3 days. Store in the dark in glass jars with tight-fitting lids.

RACK DRYING
This is the way to proceed with porcini (aka ceps or penny buns), field and oyster mushrooms, fairy ring champignons, horn of plenty (which are too flimsy to thread successfully) and many other species. Larger mushrooms, such as porcini, which can be huge, should be cut into vertical slices before drying or the process will take ages.

Unsurprisingly, the first thing you need for rack drying is a suitable rack. This will allow you to spread your mushrooms so that they neither touch one another nor plummet through the gaps. If the spaces between the wires are too big, find another rack! The key to successful drying is good ventilation, so don't be tempted to lay your fungi on newspaper, which will in any case absorb some of the precious flavours. Cake racks have an appropriate fineness of mesh, though they are on the small side if you've had a bumper forage. You may need to improvise with a reel of wire. If you've made or bought a willow rack, of course, your problem is solved.

The next stage is to find a source of gentle heat. A drying box (see page 14) would be ideal, otherwise place the rack on top of a radiator or in an extremely low oven with the door ajar. Drying times will vary from species to species and with the condition of your original mushrooms, but you're probably looking at 12 hours to 24+. Even within the same batch, some individual

mushrooms will dry faster than others, so you will want to remove them as they reach the desired state. This is when they are crispy to the touch.

STORAGE AND USAGE

The quicker you process your mushrooms after picking, the better the results will be. Dried fungi look great in jars, but make sure they are airtight and kept out of direct sunlight. They will keep for up to 18 months.

Some soups and stews call for intact dried mushrooms. These will usually need to be soaked beforehand, in which case incorporate the soaking liquid in your recipe as it will have absorbed some of the flavours. Dried mushrooms can also be blended into powder. Use in soups, sauces and gravies.

DRIED PORCINI AND GRUYERE TARTS

Nick tends to use his best dried porcini for this tart, but it could be made with nutty morels or dried fairy ring champignons. Sometimes, he sets such tarts in his hot smoker (see chapter 3). **Serves 4**

THE PIE SHELL
225g (8oz) plain flour
½ teaspoon salt
1 teaspoon sugar
100g (3½oz) unsalted
  butter, cut into small
  pieces

THE FILLING
200g (7oz) mushrooms
  (e.g. chestnuts – small
  Portobellos), sliced
50g (2oz) unsalted butter
25g (1oz) dried porcini
  mushrooms (see page
  21), dirt free, cut into
  small pieces
2 shallots, sliced
1 head of garlic, roasted
  as per page 127
50ml (2fl oz) dry sherry
100g (3½oz) crème fraîche
125g (4½oz) Gruyère
  cheese, grated
5 medium eggs
Paprika
Salt and pepper

- To make the pastry, sift all the dry ingredients into a bowl and then, using the tips of your fingers, rub the butter into the flour until it is granular and airy. Add water until the ingredients press together into firm dough. Cover with clingfilm and leave in the fridge until you need to roll it out.
- Preheat the oven to 180°C/350°F/gas mark 4. Roll out the pastry and transfer to a 20–25cm (8–10in) pie dish. Press the pastry into the dish and fold the overlap over the lip so that it hangs down.
- Prick the pastry with a fork, line it with baking paper and weigh it down with dry pulses or rice.
- Bake for 15 minutes, then remove the paper and the pulses or rice. Turn up the heat a notch to 190°C/375°F/gas mark 5 and bake the pie shell for a further 15 minutes until lightly coloured. Cut off the overlapping pastry with a sharp knife.
- To make the filling, fry the fresh mushrooms in the butter over medium heat with the dried porcini, shallots and roasted garlic for a few minutes. Add the sherry and heat until it has almost evaporated.
- In a suitable bowl, thoroughly mix the crème fraîche, Gruyère, eggs, a little salt and pepper and the mushrooms.
- Spoon the mixture into the pie shell and sprinkle with paprika. Bake in the oven for 30–40 minutes at 180°C/350°F/gas mark 4 until set. Or hot-smoke for 1 hour at 100°C/212°F.

PORK AND DRIED SHIITAKE MUSHROOM SOUP WITH DRIED ANCHOVIES

Dried shiitake are arguably better than fresh ones, with a deep woody flavour and velvety chewiness. If you find yourself hooked on the fungi, you can buy logs pre-innoculated with the spore and grow them at home. Dried anchovies are salty and pungent, but they do whiff a bit if you prepare them at home, and the neighbours may not thank you. Fortunately, they are readily available in oriental supermarkets. This nutritious, full-flavoured soup makes excellent use of both ingredients. **Serves 4**

400g (14oz) pork fillets marinated in the juice of 1 lime, 1 heaped teaspoon honey and 1 tablespoon dark soy sauce

2 litres (3½pints) chicken and pork stock (see method for ingredients)

2 small handfuls of dried anchovies

100ml (3½fl oz) vegetable oil, plus a little extra

10 dried shiitake mushrooms

Sesame oil

2 teaspoons sliced ginger

3 medium-strength dried red chillies (see page 26)

300g (10½oz) mini pak choi or greens of your choice such as spinach or broccoli

Rice noodles

Soy sauce

A splash of rice wine

2 teaspoons sesame seeds, dry fried for 2 minutes over a moderate heat until slightly browned

- Start marinating the pork fillet several hours before you plan on eating it.
- To make 2 litres (3½ pints) chicken and pork stock, take 2kg (4½lb) chicken and pork bones and offcuts and bake them for 40 minutes at 220°C/425°F/gas mark 7. Then empty them into a large saucepan, remembering to scrape in all the juices. Add 2 medium carrots, roughly chopped, 1 medium onion, cut in half, 4 cloves of garlic, cut in half, and 2 whole star anise. Fill with water until the solid ingredients are slightly more than covered. Simmer for 2½ hours, skimming the fat off the top at regular intervals, then strain and reserve. If you find that you don't have quite enough stock for the recipe, top up with water.
- Make sure the anchovies are gutted and headless, or the flavour will be tainted. Fry the fish in the vegetable oil over moderate heat until golden, then place them on kitchen paper to soak up the oil. Reserve until required.
- Cover the dried shiitake with boiling stock, then leave to soak for 1 hour. You can slice them now if you wish.

- Fry the pork in a little sesame and vegetable oil over moderate heat until nicely browned on all sides and cooked through. Better still, cook on the char grill.
- Heat the stock in a large saucepan, then add half the anchovies. Simmer for 5 minutes, then take out the fish with a slotted spoon and discard. Add the mushrooms, ginger and chillies and simmer on for 5 minutes.
- In the meantime, blanch the pak choi or other greens in boiling water for a couple of minutes, then plunge them into cold water. This will stop the vegetables cooking and keep them bright green.
- At this stage, you should also prepare your rice noodles according to the instructions on the packet.
- Season the soup with soy sauce and rice wine, then add the pak choi or other greens.
- Pour over the noodles, then slice the pork and place on top. Garnish with the rest of the anchovies and the sesame seeds.

DRIED CHILLIES

In New Mexico in the south-western United States, chillies are treated with the same reverence as wine among French connoisseurs. The relative merits of different varieties are debated with ferocious passion. Spiky wreaths or ropes of dried chillies called *ristras* hang everywhere. Aside from providing readily available handfuls of the local staple, they also look extremely festive. In recent years, Northern Europeans have cottoned onto this in a big way. In chic boutiques, Christmas trees are now just as likely to be adorned with deep-burgundy dried chillies as more traditional baubles.

Most kinds of chillies can be successfully dried at home, with the exception of thick-walled, meaty varieties like jalapeños. These are nonetheless excellent when smoke-dried, whereupon they become chipotles (see instructions opposite).

For other species, you will get the best results if you stick to fully ripened chillies, in other words red ones. Your chillies are fully dried when they snap rather than just bend. They will remain in good condition for up to a year.

RACK DRYING

The easiest and most effective way of rack drying chillies is to use a food dehydrator, and we strongly advise you to invest in one if you are catching the drying bug (see page 218 for suppliers). The next best approach is to use a home-made drying box, which we showed you how to make on page 14. The third option is to use an oven turned to its lowest setting with the door left slightly open. This is not particularly economical and may annoy your significant others if they want to cook something else, but at least you don't need any specialist equipment.

If you are using a dehydrator, simply follow the instructions. With the other two methods, it is first advisable to halve the chillies lengthwise and remove the seeds (and then watch what you do with your hands for a while!). For oven drying, you should aim for a temperature of 60–70°C/140–160°F and expect to wait 24 hours or more. The drying-box technique will probably take twice as long, but it's less disruptive to your domestic routine.

STRING DRYING/RISTRAS

The easiest approach is to pass a threaded needle through the base of the stems of a succession of chillies until you have a long line of them. But if you are feeling creative, you may want to construct a fully fledged *ristra*. This involves tying chillies together in clusters of three and braiding them along a length of wire. New Mexico chillies are the optimum variety to use. Here's how to go about it:

1) To make a 45cm (18in) *ristra* you will need about 3.5 kg (7–8lb) fresh chilli pods and several 1.5m (5ft) lengths of cotton string.
2) Take three chillies and hold them together by their stems. Wrap the string around the stems twice, then pass it down and then up again between two of the chillies and pull tight so it cuts into the stems slightly.
3) Loop the string around the hand holding the stems, then move the loop so that it rests over the ends of the stems. The free end should hang down through the loop. Pull tightly. Scouts, guides and sailors will recognise that they have just tied a half-hitch.

4) Carry on making clusters every 7.5cm (3in) along the string. Continue until all the chillies have been used, starting new strings as necessary.
5) Suspend a length of wire or a straightened coat-hanger from a nail, door knob or other convenient place. Make a small loop at the bottom to prevent the chilli clusters from slipping off.
6) Starting with the first cluster of chillies on the end of one of your strings, braid them around the wire, starting at its bottom. Doing this is not unlike braiding a person's hair: the wire serves as one strand and two of the chillies in each cluster act as the other two. First twist one of chillies around the wire, then do the same with another one. Then move on to the next cluster and so on. When you finish braiding one string of chillies, move on to the next one until the *ristra* is finished. Make sure you vary the direction in which the braided chillies point out from the wire to guarantee a nicely three-dimensional *ristra*. And as you finish braiding each cluster, push it down in the centre to ensure that the chillies are densely packed.

Whether you've made a simple string or a complicated *ristra*, the next stage is drying. If you are lucky enough to live in a desert region, you can hang your strings of chillies in the sun and wait for them to dry, though bring them in at night to avoid dew. Otherwise, you will need to use one of the techniques listed in the rack-drying section above. Remember to ruffle the chillies up and turn them from time to time to ensure even drying. In moist old Britain, we tend to hang them in the airing cupboard for a week afterwards to make sure they are completely dry.

CHIPOTLE

Chipotle is the Mexican name for smoke-dried jalapeño. They are easy to make in the hot-smoker, particularly once you've read and digested our smoking chapter.

Take 40 red jalapeño chillies, cutting a lateral slit in each one. Lay them on a rack in the hot smoker and smoke for 3–4 hours at 100°C/212°F. At this point they will be somewhat dehydrated but not completely desiccated. Finish them off in a cool oven (80–90°C/180–200°F/gas mark ¼) until dry. Chipotles impart a wonderful smoky flavour to stews and sauces and are the key ingredient in adobo seasoning (see page 130).

TOMATOES

In recent times, sun- and semi-dried tomatoes have become indispensible to cooks wherever they happen to live. But unless you can reliably predict several days of breezy weather with low humidity and daytime temperatures in excess of 32°C/90°F, which is roughly never where we live, you'll have to fall back on other methods. As usual, the chief options are using a dehydrator, a low oven with the door ajar or your home-made drying box. An ingenious alternative is to place a rack of tomatoes on the shelf under your car's rear window on a hot day.

FULLY DRIED TOMATOES

Drying times will vary according to the size of your tomatoes, but as a rule of thumb, 15 hours in a low oven or 30 in a drying box is about right. However, tomatoes in any given batch will not dry at exactly the same rate, so you need to remove them individually as they become ready. This is when they are firm but no longer juicy.

Whichever method you use, you have two main choices. The first is to cut the tomatoes in half and lay them face up on a fine-meshed rack, sprinkling a few grains of sea salt on each face. The second is to dry them intact on the vine. This involves lying the tomatoes on a similar rack, vine stalk down, before cutting a small cross on the top of each and filling it with a pinch of salt.

Once dried, tomatoes can be stored at ambient temperatures in sealable containers for up to 6 months. Before use, they will need to be rehydrated by soaking in warm water for half an hour. They should always be cooked before they are eaten.

SEMI-DRIED TOMATOES

As the name suggests, semi-dried tomatoes are removed from the source of heat half way through the drying process. They are then packed into sterilised pots (see page 164) which are filled with olive oil. These will keep in the fridge for up to 6 months. They are moist and more than good enough to incorporate in stews, sandwiches and sauces without further ado. We've achieved our best results using various varieties of cherry tomatoes.

PISSALADIERE WITH SEMI-DRIED TOMATOES

Pissaladière is the Provençal version of pizza. Whereas the Italians tend to use concentrated tomato sauce as the base flavouring, in the south of France they are as likely to use slices of onion slow-cooked in olive oil. This recipe, which incorporates semi-dried tomatoes, gives you the best of both worlds. At the market in St Tropez, many kinds of pissaladière are on sale, topped with various permutations of courgettes, goat's cheese, pine nuts and red peppers. **Serves 2–3**

THE DOUGH

450g (1lb) '00' grade flour

2 teaspoons salt

1 teaspoon dried yeast

2 tablespoons olive oil

300ml (½ pint) water, hand-hot

THE TOPPING

1kg (2¼lb) onions, sliced

100ml (3½fl oz) olive oil

A few small sprigs of fresh rosemary and thyme or a pinch or two of dried

15 anchovies, if salted (see page 45), soak before use; if preserved in oil they are ready to use

15–20 small black olives

20–30 semi-dried cherry tomato halves (see page 28)

100g (3½oz) crumbly goat's cheese (optional)

- First, make the dough. Doing this by hand is fun but you could just as easily throw it together in a food-processor.
- Begin by sifting the flour into a large bowl. Mix in the salt and dried yeast, then stir in the olive oil and water with a large spoon.
- Turn the sticky mass out onto a floured surface and knead until it becomes smooth. Or do what you need to get the same result from a food-processor.
- Return the dough to the bowl and cover with a damp cloth. Put the bowl somewhere warm (an airing cupboard will do fine, or any available space in a warm kitchen) and wait for an hour for the dough to rise.
- When this happens, knead it for a minute, then divide it into two balls. You will need just one of them for the pissaladière. The other one can go in the freezer.
- To make the topping, first sweat the onions in the olive oil for 30–40 minutes, along with the rosemary and thyme. Stir frequently and keep the heat low. At the end, they should be soft and sweet, but not browned.
- Preheat the oven to 230°C/450°F/gas mark 8. Roll out the pizza dough gradually, using a little flour to prevent it from sticking. Nick rolls his directly onto a rectangular baking sheet that slots straight into his oven. Alternatively, you could use a pizza stone or rectangular baking tray, or place the pissaladière naked on the shelf of a wood-burning oven.
- Before adding the topping, turn up the edges of the dough by pinching between your thumb and forefinger.
- Now spread the onion mix evenly over the dough. Arrange the anchovies in a lattice and sprinkle with the olives, tomatoes and goat's cheese, if using.
- Place in the hottest part of the oven and cook for around 20 minutes. Check after 15 minutes, because ovens are variable in their accuracy. Consume immediately.

FRUIT

When you dry fruit, the flavours intensify and the natural sugars become more concentrated, so it makes sense to use perfectly ripe specimens. Sun drying is great if you can be confident of four or five days of hot, sunny weather, but many of us are seldom that lucky. Otherwise, the best drying vehicle is a multi-tiered dehydrator, though as usual a very low oven will do fine, as will a drying box (see page 14). As far as racks are concerned, you can either use the regular wire kind or weave your own from willow switches. For smaller fruit, you may want to cover your rack with scalded muslin or cheesecloth – the scalding is important to keep the material from scorching or passing on its taste to the fruit.

The key to successful drying is that the process should be slow and steady. Too rapid and the fruit will become tough and wrinkled and may split. Too slow and it may rot. It is essential that every fruit in a batch is properly dried before storage. If one remains moist and goes bad, the trouble will spread to all the others. Your fruit is ready when it feels leathery and releases no moisture when squeezed.

Dried fruits can be eaten as they are or else they can be reconstituted by soaking them in lukewarm water or wine for about 24 hours.

SOFT FRUIT
Because of their high water content, soft fruits take longer to dry than hard ones. It is important to keep the temperature below 70°C/160°F, at least for the first hour, to prevent the surfaces from hardening, as this would hinder evaporation.

Plums and Damsons
Dried plums are of course prunes. Don't let painful memories of school puddings put you off – home-prepared prunes are in another league entirely.

You can slice your plums before drying them, halve them, or cut slits in them and remove the stones. Whole plums take longer to dry, perhaps as much as 2 or 3 days. Slices should be ready in 12 to 24 hours. We sometimes add a little sugar before drying, but this isn't necessary. We just like 'em sweet.

Smoked prunes are also very good. Hot-smoke them for 12 hours at 100°C/212°F, preferably using cherry, apple or plum wood, and finish them off in the oven at its lowest setting.

Damsons, which are wild relatives of the plum, can be dried in the same manner as their domesticated cousins.

Nectarines and Peaches
Treat these in the same way as plums, although they are less appealing when smoked.

Figs
Slice and sprinkle lightly with sugar, then dry as per peaches.

Grapes
Use seedless red or black varieties. Blanch in boiling water for a few seconds, then refresh under cold water to help break up the skin and thus speed up the dehydration process. Lay the grapes out on a very fine-meshed rack, otherwise they will fall through. They take rather a long time to dry, about 48 hours in a low oven (80°C/176°F), but you can't beat home-made currants and raisins.

Blueberries and Cranberries
Use the same method for blueberries as for grapes. Ditto with cranberries, although we like to dip them in a solution of 1 tablespoon of honey for every 250ml (9fl oz) of water before drying them to counteract their tartness.

Bananas

Cut your 'nanas into slices approximately 1cm (½in) thick and soak them in a solution of 30 per cent lemon juice and 70 per cent water for 3 minutes. The citric acid in the lemon juice will prevent later discolouration. They will be nice and crispy after about 20 hours in a low oven (80°C/176°F/gas mark ¼).

Strawberries

Slice them about 5mm (¼in) thick and lay out in a single layer on a drying rack. Turn over once during drying, which should take about 12 hours.

Kiwi Fruit

Simply cut kiwis into slices and dry for around 12 hours.

Pineapples

As per kiwi fruit, only they take longer to dry.

Cherries

These need to be pitted before drying and placed in a single layer on a fine-meshed rack. They will be ready in 18–24 hours.

HARD FRUIT

Dried hard fruits make excellent snacks and are useful as supplements for home-made muesli.

Apples and Pears

To make crunchy rings of apple or pear, you first core the fruit with a fruit de-corer (!). Then you decide whether you want to skin them or not (we tend to leave the skins on) and take the appropriate measures if the former. Next, slice them into rounds approximately 5mm (¼in) thick and dip these for 3 minutes in a solution made up of 100ml (3½fl oz) lemon juice to 200ml (7 fl oz) water to 1 teaspoon sugar. Then dry the rounds in a dehydrator, low oven or even a microwave, or over a warm radiator. This should take 12–18 hours if you're using an oven. A single layer of apple or pear rings can be dried in 3–5 minutes in a microwave set to 'defrost'.

DRIED FIG AND PRUNE MILLEFEUILLE

If you make this luxurious creamy dessert once autumn has set in, you'll be glad you bothered to make the effort to dry all those figs and plums at the end of the summer. You can either buy the *langue de chat* ('cat's tongue') biscuits in the supermarket, or better still make them yourself. **Serves 4**

THE LANGUES DE CHAT
100g (3½oz) butter
100g (3½oz) icing sugar
2 medium eggs, lightly
 beaten
125g (4½oz) plain flour

THE PRUNE FILLING
100g (3½oz) stoneless
 prunes
300ml (½pint) port
2 tablespoons sugar

40 dried fig slices (see
 page 32)
200ml (7fl oz) thick double
 or clotted cream

- To make the *langues de chat*, preheat the oven to 190°C/375°F/gas mark 5.
- Cream the butter and icing sugar together and whisk for a good few minutes until pale.
- Slowly add the eggs and flour, whisking constantly.
- Pipe short lines of the mixture, each approximately 6cm (2½in), onto a greased baking tray. Place in the oven and bake for 8 minutes or until slightly coloured. Leave to cool and transfer into an airtight container. This can all be done well in advance of making the millefeuille.
- To make the prune filling, simmer the prunes in a pan with the port and the sugar until the volume has reduced by at least

half. Blend the mixture in a food-processor until smooth and reserve.

- To construct each serving, do as follows: place 2 *langues de chat* side by side on a dessert plate. Spread a layer of prune purée and cream over the biscuits. Then add a layer of dried fig. Then another layer of purée and cream, and finally some more *langues de chat*. Scatter dried fig slices around the plate, and serve with double cream or custard. Alternatively, make up some Chantilly cream by adding 2 tablespoons of sugar and ½ a teaspoon of vanilla extract (or the seeds from half a vanilla pod) to 225ml (8fl oz) whipping cream. Whisk until soft peaks form.

SEA BASS WITH DRIED DAMSONS

One satisfying and often very successful way to devise a 'new' dish is to take a foreign classic and adapt it to suit local ingredients. Sea Bass with Dried Damsons is a case in point. In China, this would be made with pickled plums. These have a startlingly intense flavour, but some Westerners find them unbearably sour. This is a pity, because the underlying taste combination is a winner. The solution turns out to be to use dried damsons. These wild relatives of the plum are every bit as sharp and flavoursome as their oriental equivalents, but sweeter. And round where we come from, they're pretty much free. **Serves 2**

10 dried damsons (see page 32) or use 3 pickled plums

1 sea bass (about 600–800g/1¼–1¾lb), trimmed, scaled and gutted

3 spring onions, sliced

2 teaspoons chopped ginger

3 medium-strength dried red chillies (see page 26)

Juice of ½ a lime

Splash of soy sauce

Generous splash of fish sauce

1 teaspoon chopped garlic

1 teaspoon sesame seeds

2 tablespoons sesame oil

1 tablespoon vegetable oil

- Immerse the damsons in enough boiling water to cover. Soak for 1 hour, then remove the stones and roughly chop.
- Place the sea bass on the largest flat plate that you can fit in your steamer. If you have to chop the head off, so be it.
- Sprinkle all the remaining ingredients over the bass with the exception of 1 tablespoon of the sesame oil and the vegetable oil.
- Steam the fish for 12–15 minutes. Sea bass doesn't take much cooking. Gently pierce the flesh through to the bone to see whether it is done. A clear, dark amber sauce will have collected on the plate.
- To finish this dish off heat the remaining oils in a pan until almost smoking, then pour over the fish, which will fizz up dramatically.
- Serve with plain rice and greens with oyster sauce.

VENISON WITH DRIED CRANBERRIES

The foundation of this dish is the beef demi-glace which can be frozen successfully, so don't worry if you make too much. The cranberries are added right at the end. Serves 4

THE DEMI-GLACE

2kg (4½lb) beef bones, cut into small pieces

500g (1lb 2oz) veal bones, cut into small pieces

4 cloves of garlic

2 carrots

50g (2oz) fresh mushrooms

1 large onion, cut in half

3 sticks of celery

100g (3½oz) leek

80g (4oz) wheat flour

A sprig of thyme

160g (5½oz) tomato purée

A sprig of parsley

A sprig of rosemary

4 bay leaves

10g (½oz) dried porcini (see page 21)

½ teaspoon peppercorns

salt

100ml (3½fl oz) port

THE MAIN DISH

125g (4½oz) dried cranberries (see page 32)

About 450g (1lb) venison strip loin

Salt and pepper

A little olive oil

Fresh dill, to garnish

- To make the demi-glace, preheat the oven to 200°C/400°F/gas mark 6 and roast the beef bones, veal bones, garlic, carrot, fresh mushrooms, onion, celery and leek in an oven dish for 1 hour or until lightly browned. Sprinkle the flour on top and roast for a further 10 minutes.
- Transfer the contents to a large saucepan and swish out the bottom of the oven dish with boiling water. Scrape off the juicy bits and pour them into the saucepan. Add the thyme, tomato purée, parsley, rosemary, bay leaf, porcini and peppercorns.
- Top up with water to around 4cm (1½in) above the bones and bring to the boil. Simmer for at least 8 hours. Regularly skim the surface of the demi-glace for fat and impurities.
- Pass the demi-glace through a conical sieve into a clean pan. Reduce to around 1.5 litres (2½ pints) by gently boiling and season with salt.
- Portion off 1 litre (1¾ pints) of the demi-glace and freeze this.
- Reduce the port by half in a saucepan and then add the remaining 500 litre (18fl oz) demi-glace. Add half the cranberries and simmer for 40 minutes. Pass through a conical sieve, pressing the cranberries firmly as you do so to smash them. Then add the rest of the cranberries to the collected juice and simmer while you cook the venison.
- Season the venison and dab the meat with a little olive oil. Sear in a frying pan on each side, then place on a very low heat and continue to cook until rare, medium or however else you like it.
- Spoon a little demi-glace onto each plate with a few cranberries and then lay slices of venison on top. Garnish with the fresh dill and serve.

FRUIT LEATHER

Fruit leather is essentially concentrated fruit, puréed, dried and rolled into sheets. The colours are vivid and the flavours intense. Kids love fruit leather and you have the comfort of knowing they are snacking on something healthy.

You can have fun experimenting with leathers made from a variety of fruits and combinations thereof. We give you recipes for a couple of winning formulas below. The first is lovely and yellow, the second lovely and purple. Just remember to add lemon juice if you are using fruits that are liable to discolour. If you overcook your leather, it becomes brittle and difficult to peel off the aluminium foil or clingfilm. The same applies if you add too much honey or sugar.

MANGO AND YELLOW PLUM FRUIT LEATHER

Makes 1 sheet

1 medium mango, ripe and sweet
4 medium yellow plums
80ml (3 and a bit fl oz) honey

- Peel and dice the mango and do the same with the yellow plums.
- Place the fruit in a saucepan with the honey and simmer for 5 minutes.
- Blend with a hand-blender until smooth.
- Line a medium-sized rectangular baking tray with clingfilm or aluminium foil and pour in the mix until it just runs to the sides.
- Place in an oven heated to 70°C/160°F/gas mark ¼ and leave for about 6 hours. The leather is ready when it is tacky but no longer sticky.
- Leave the leather to cool, then roll it up in clingfilm or cut it into strips and store them in an airtight container in a cool place. It will keep for 2 months at ambient temperature in a dark place, 4 months in the fridge or 1 year in the freezer.

FRUITS OF THE FOREST FRUIT LEATHER

Makes 2 sheets

200g (7oz) raspberries
300g (10½oz) strawberries
150g (5oz) blueberries
Juice of 1 lemon
100g (3½oz) honey

- Boil up the raspberries in a pan, then pass through a food mill or conical sieve. Keep the juice and discard the seeds. Return the raspberry purée to the pan and add the rest of the fruit. Add the lemon juice and honey and simmer for 5 minutes.
- Blend with a hand-blender until smooth.
- Line a medium-sized baking tray with clingfilm or aluminium foil and pour in the mix until it just runs to the sides.

- Place in an oven heated to 70°C/160°F/gas mark ¼ and leave for about 6 hours. The leather is ready when it is tacky but no longer sticky.
- Leave the leather to cool, then roll it up in clingfilm or cut it into strips and store them in an airtight container in a cool place. For storage times, see Mango and Yellow Plum Fruit Leather recipe on page 38.

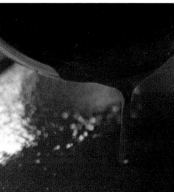

DRIED SHRIMP

To amuse his young son, Nick recently bought a shrimp net in Cornwall from a man who seemed to have seen him coming. 'There won't be any shrimp there daddy!,' cried Jack as his father led them towards a pungent inlet. Nevertheless, they slid down the slimy slope and Nick dipped the net into the stagnant water. He poked it around under some rocks in hope rather than expectation. When he withdrew it, they craned their necks to inspect the trawl. Hundreds of plump shrimp erupted into their faces. Jack is still having therapy. Nick has a never-ending supply of dried shrimp.

For the benefit of American readers, in Britspeak the term 'shrimp' refers to very small crustaceans indeed. They don't taste of much in their fresh state, but no one could level this accusation at them once they've been dried. Left whole, they go nicely in a Mediterranean fish stew. In this case, you will need to rehydrate them in warm water for about 12 hours. Blended into a powder, they are an integral ingredient of Pad Thai.

TO DRY SHRIMP

Your shrimps should be very small, i.e. 200 or more to the kilo (or 90 plus to the pound). Blanch them in boiling water for 3 minutes, then lay them out to cool. Take the heads off and discard them.

If you are planning to use your shrimps whole, peel them before drying. If you are intending to powder them, there's no need to bother.

Dry on a wire mesh in a low oven (80°C/175°F/gas mark ¼) for 12 hours or until completely dry. Store in an airtight container for up to 6 months in a dark larder or 1 year in the fridge.

PAD THAI WITH SHRIMP

This is what travellers to Thailand crave when they get back to their home countries. Pad Thai is also popular with people who have scarcely heard of the place. Dried shrimps add authenticity, flavour and crunch. Serves 3–4

2 tablespoons blanched peanuts
Soy sauce
Honey
250g (9oz) minced pork
1 tablespoon sesame oil
2 tablespoons vegetable oil
8 baby corn, sliced thinly
½ red pepper, thinly sliced
10 fresh shiitake mushrooms, thinly sliced
1 red chilli, finely chopped
1 cube of ginger, chopped
2 cloves of garlic, chopped

2 shallots, thinly sliced
2 tablespoons small dried shrimp (see left)
2 heaped tablespoons smooth peanut butter
Juice of 1 lime
180g (6½oz) shredded cabbage, blanched briefly in boiling water, then plunged into cold water
500g (1lb 2oz) rice noodles (cooked weight)
Fish sauce
3 spring onions, sliced

- Coat the peanuts in a mixture of soy sauce and honey, and bake for 10 minutes at 190°C/375°F/gas mark 5 or until lightly browned
- In a large wok, fry the pork over quite a fierce heat, stirring and tossing constantly. Remove from the pan and reserve. Then add the sesame and vegetable oils and fry the baby corn, red pepper and shiitake mushrooms for 5 minutes until they start to colour.
- Add the chilli, ginger, garlic, shallots and dried shrimp and continue to fry over a moderate heat for another 5–8 minutes, stirring constantly.
- Return the minced pork to the wok along with the peanut butter, lime juice, shredded cabbage and noodles (cook them first). Toss until combined. Season with soy and fish sauce. Garnish with spring onion and honey-roast peanuts.

SALTING

If you are lucky enough to attend a sumo wrestling bout in Japan, you will see enormous men in loincloths cast handfuls of salt over the ring before they charge at each other with the momentum of mini-elephants. The idea is to ward off evil spirits. Salt is associated with purity, and evil spirits can't handle that.

A certain amount of salt is necessary for effective metabolism, which is why blood tastes salty. But the real reason why small rocks of sodium chloride have played such an important role in human history lies elsewhere, in the miracle of osmosis. This is the compulsion felt by liquids enclosed by semi-permeable membranes, such as the walls of plant or animal cells, to get themselves in sync with whatever lies on the other side. If a soluble substance like salt is rubbed on a piece of fish, for example, the cells inside leach water in an attempt to dilute their surroundings until they are no more salty than the original cell contents. Hence a cucumber rubbed with salt will ooze prodigiously. Of course, if fresh food is packed in salt or immersed in a large tub of concentrated brine, its cells are fighting a losing battle, and it will ultimately become very dry indeed. This makes it unappetising to bacteria, which are in any case put off by salt water because, on the same osmotic principle, it tends to dry them out. This is why the Dead Sea, which has a salt content of around 30 per cent, remains so resolutely dead.

All this helps explain why, until relatively recently, salt has been such a precious commodity that wars have been fought over it. Its power to drain foods of their moisture renders them useless for bacteria, but very handy for us. The historical importance of salt is demonstrated by the fact that Roman soldiers were sometimes paid in it, a practice that gave us the word 'salary'. Particularly in damp countries where conditions were not ideal for drying, salt was often the sole way of preserving enough food to keep our ancestors alive during the winter. The killing and salting of the family pig in autumn was a ritual on which millions of European and American peasants depended. And Gandhi, remember, kicked off his peaceful revolution by daring to make salt, a lucrative procedure which the British had hitherto jealously monopolised.

With the advent of refrigeration, salt has lost some, but only some, of its centrality to our lives. Only now can we afford the luxury of worrying about the amount of sodium in our diets (although the traditional Lenten fast may have fulfilled a similar function, being a prudent precaution in the days when, by springtime, everyone would have been guzzling salty meat for months). We have come to regard salt primarily as a seasoning. But it remains vital to a surprising number of preserving processes. It also has another advantage: by concentrating the flavour of food through depriving it of tasteless water, it makes it thoroughly delicious.

DRY-SALTED ANCHOVIES

The anchovy, a tiny silver fish which swims in huge schools in the temperate waters of the Mediterranean and off the coasts of Peru and California, has played a major role in the history of preserving. This is partly because, despite its abundance, its delicate white flesh deteriorates so quickly when exposed to the air that few people apart from fishermen ever get the chance to eat it fresh. Instead, it is salted in industrial quantities and canned or bottled in either pure salt or olive oil.

The Romans were addicted to anchovies, particularly in the form of *garum*. This fermented condiment, the ancestor of Worcestershire sauce and cousin of the great South-East Asian fish sauces, was made from the liquid drained from barrels of curing anchovies. The Italians, particularly in the south, still use anchovies in a range of contexts that might seem bizarre to foreigners – for instance, with roast lamb. Across the pond, meanwhile, they are a vital ingredient in Caesar salad, the name of which inadvertently pays tribute to an ancient love affair.

Salted anchovies are 'meatier' than their equivalents in olive oil. You will need to soak them in cold water for 5 minutes or so before using them, and to remove the backbones and tails, but it is well worth it.

TO SALT ANCHOVIES
1 kg (2¼lb) fresh anchovies
1 kg (2¼lb) rock salt

Clean and gut the anchovies.

Cover the bottom of a terracotta jar with rock salt, then alternate layers of anchovies with layers of salt until the jar is full, finishing with a layer of salt.

Place a wooden disc inside the jar on top of the last layer – you may need to do a bit of carpentry to obtain one the right size. Then weigh it down with something appropriate. The weight will cause the excess liquid to rise to the top. You should remove it with a clean cloth from time to time. The anchovies will be ready after 2 to 3 weeks, by which time the salt will have given them an almost 'cooked' texture. If you are able to refrain from eating them all straight away, pack them into small, sterilised jars (see page 164) with either more salt or some good olive oil and keep them in the fridge (see page 46).

DRY-SALTED ANCHOVIES

SALTED CAPERS

SALT COD

There is no point pretending that the board-hard slabs of salt cod sold in Mediterranean food stores and certain fishmongers are particularly prepossessing. But when, after what seems like an eternity of soaking and rinsing, you have cooked them up, perhaps in the Spanish style with onions, tomatoes and garlic, you will find the taste and texture of the fish transformed compared with those of the fresh version. Salt cod also soaks up the flavours of the ingredients it is cooked with.

Properly prepared salt cod does not actually taste salty. It has an intense taste that somehow lies 'beneath' the surface, and a firm, chewy consistency that is more reminiscent of meat than of fish. Historical necessity has hooked wide swathes of the world's population on salted cod, from the Jamaicans whose national dish is saltfish and ackee (a strange fruit which comes out like scrambled eggs when cooked) to the Portuguese who are lost without their weekly dose of *bacalhua*. To the Spanish, salt cod is *bacalao*, while to the Italians it is *baccalà*. Around Naples, Norwegian lorries are still regularly hijacked to feed the local appetite for the delicacy. Not all those who weren't weaned on the habit take to salt cod at once, but those who do like it very much, as do the millions who have a taste for it inscribed in their genes.

Industrially produced salt cod is cured in huge vats of salt for about 2 weeks, before being dried in a 'hot room' for 24 hours. If you decide to make it yourself, you can skip the drying stage, as this is more about ease and economy of transportation than anything else.

TO MAKE SALT COD

Get hold of several fillets of the Atlantic variety and a large sack of coarse sea salt. You then proceed much as you do with the anchovies (see page 45).

Take a large container made of an unreactive material like porcelain or white plastic and pour a layer of salt into its base. Then add a layer of fish, pressing it into the salt below, then another layer of salt, then fish again and so on, finishing with a top layer of salt. The cod will be ready in 3 days, but you should leave it in the salt for at least a week to 10 days depending on its thickness. After a day or so, break up any crusts of salt that have appeared to prevent air pockets forming around the fish. It needs to be in direct contact with the salt throughout the process.

When you want to use the fish, remove the amount you need from the salt, rinse it thoroughly and soak it in fresh water for 2 or 3 days, changing the water at least twice a day. Bear in mind that it will probably double in size during the soaking phase.

Store your salt cod somewhere cool, for instance, in a shed or cellar. It will keep for at least 1 year.

SALT PORK

When reading accounts of peasants or sailors from times past subsisting on salt pork for months on end, it is easy to feel sorry for them. But they should be pitied for the monotony of their diet rather than the fact of it, for this most fundamental of preserved meats is actually rather delicious. A few scrapings of flavoursome salt pork can transform the least promising stew or gruel into an enjoyable dish, as spotted by populations as diverse as the Chinese and the Jamaicans.

Salt pork is usually made from pork belly and often from the fattier portions thereof. It can be either dry-cured or brined. We prefer the dry-curing method, and here's how we do it:

TO MAKE SALT PORK

Take 5kg (11lb) of pork belly and cut into strips of about 1kg (2¼lb).

Trim off the bones and make regular incisions in the skin.

Get hold of a wooden box and sprinkle a layer of sea salt at the bottom.

Lay a single layer of pork on top, cover it with salt and massage it in a little. Repeat the process until all the pork is covered with salt, leaving a layer of salt on the top.

Place a lid on the box and leave in a cool place for 4 weeks. The juices will run from the base of the box as the salt draws them out by osmosis, so you will want to place a suitable receptacle underneath to catch them.

Every few days check it to see that the pork is well covered with salt. If it isn't, don't be shy about adding some more.

When you remove the pork from the box, wipe it down with a cloth, then wrap it in muslin or in a paper bag and hang it in a cool, dark, airy place. It will keep for years, though it will become harder as time goes by. You may want to halt the hardening process by transferring the pork to an airtight container and placing it in the fridge after a month or two.

If it is very hard, salt pork needs to be soaked before you cook with it. You may want to soak or at least rinse it thoroughly anyway, as it is unsurprisingly on the salty side.

Salt pork is delicious chopped up into small pieces and fried until crispy. It goes very well with rice, peas and salt cod.

PORK 'N' FISH 'N' NOODLES

When it comes to simple and delicious Asian cooking, this dish is the bee's knees. Nick's son Jack loves it because it contains goodly amounts of salty crispy bits. And because his old man cures his own pork and dries his own fish, the raw materials are always to hand. **Serves 2 adults or lots of children**

THE TOPPING
3 shallots
4 cloves of garlic
Salt
A little corn flour
150ml (¼ pint) veg oil

THE MAIN DISH
50g (2oz) dried fish, cut into small dice (Nick uses home-dried mackerel, but you can buy various varieties of dried fish fillet in Asian supermarkets. Spanish mackerel is good)
50g (2oz) salted pork belly, cut into small dice (see page 49)
1 tablespoon vegetable oil
2 shallots, thinly sliced
1 clove of garlic, chopped
300g (10½oz) wide rice noodles (cooked weight)
100g (3½oz) cucumber, grated
Juice of 1 lime
Tamari

- To make the crispy topping, slice the shallots and garlic thinly, lightly season with salt and a couple of pinches of corn flour, then fry in the oil over moderate heat for 3–5 minutes or until crispy. Drain on kitchen paper and reserve.
- For the main dish, fry the dried fish and diced pork in 1 tablespoon of vegetable oil over low to moderate heat, stirring constantly until golden and crispy. Remove with a slotted spoon and reserve.
- Using the same oil, fry the shallots and chopped garlic over a moderate heat for a couple of minutes, then stir in the noodles. As soon as they are hot, add the pork and fish, cucumber and lime juice. Season with tamari.
- Serve in bowls and garnish with the crispy garlic and shallot.

BACALAO AND POTATO FISH CAKES

Nick once purchased a hunk of top-quality bacalao in a rambling Valencian market. 'Te conozco bacalao aunque vayas disfrazado', the suave vendor kept murmuring. All the old ladies in the queue tittered and gazed up at him amorously. But he was saying nothing more saucy than 'you will always know it's *bacalao*, even in disguise'. It must have been the way he said it. **Makes 8 fish cakes**

2 shallots, finely diced
2 cloves of garlic, finely chopped
200g (7oz) top-quality bacalao fillet (see page 48), soaked in water in the fridge for 24–48 hours, drained and roughly chopped
1 tablespoon olive oil
25g (1oz) butter
450g (1lb) floury potatoes, boiled until soft
1 teaspoon picante pimenton (ground hot paprika)
¾ teaspoon salt
Ground black pepper
1 large egg
A sprig of parsley, chopped
2½ tablespoons plain flour
Breadcrumbs
Oil, for frying

- Fry the shallots, garlic and bacalao in the olive oil and butter over moderate heat for 5 minutes, stirring frequently.
- Add the mixture to the boiled potatoes together with the pimenton, salt, pepper, egg, parsley and flour. Crush with a potato masher until well mixed.
- Pour the breadcrumbs onto a plate. Take about ⅛ of the fish cake mix with a large spoon and shape it into a ball. Roll this in the breadcrumbs until well coated, then flatten into a disc. Repeat until you have 8 pristine fish cakes.
- Shallow-fry each cake in oil over moderate heat for 4–5 minutes on each side until golden brown.
- Serve with roasted peppers, olive oil and lemon juice.

STIR-FRIED RICE WITH SALT FISH

When Nick's mother-in-law arrived from Kuala Lumpur recently, her luggage had a fishy aroma. 'It's my Kurau' she explained unhelpfully. 'Ah' exclaimed Susie, Nick's Malaysian nanny, 'Mah yau yee!'. It transpired that they were talking about threadfin, a fish which dries superbly. Belachan sambal is a Malaysian fermented shrimp and chilli salsa. **Serves 2**

THE BELACHAN SAMBAL
20g (¾oz) belachan (available in Asian supermarkets)
1 tablespoon vegetable oil
5 medium-strength red chillies, seeds removed and finely chopped
Juice of 1 lime
2 teaspoons sugar

THE MAIN DISH
50g (2oz) dried threadfin, Spanish mackerel or other fish, cut into small pieces

Oil, for frying
100g (3½oz) fresh tuna, cut into small chunks
500g (1lb 2oz) cooked basmati rice
4 spring onions, sliced
Leaves from a couple of sprigs of coriander
2 medium pak choi, boiled for 4 minutes in water, then chilled in cold water, drained and chopped into little pieces
Fish sauce, to season

- To make the sambal, dry-fry the belachan for a couple of minutes, then add the vegetable oil and chillies. Gently fry for another few minutes, then add the lime juice and sugar and leave to cool. The sambal can be kept in the fridge for up to 1 month.
- For the main dish, first fry 3 dried fish pieces in a little oil until crispy (about 5 minutes). Then, using the same oil, fry the tuna chunks, then add the rest of the ingredients. Stir-fry over moderate heat until the dish is steaming hot.
- Serve with the sambal on the side.

SALADE NIÇOISE

Johnny, who was an eccentric child, had a minimalist version of Salade Niçoise for breakfast every day during his primary school years. It consisted of a small tin of tuna plus a few slices of tomato and cucumber. He likes to think it enhanced his cerebral development.

As good as this youthful breakfast was, it is much more satisfying to make a full-blown version with home-salted capers and anchovies. **Serves 2**

THE DRESSING
60ml (2 fl oz) olive oil
1 tablespoon balsamic vinegar
2 teaspoons Dijon mustard
1 teaspoon honey
Juice of half a lemon
Salt and pepper

THE MAIN DISH
75g (3oz) French beans, topped and tailed
Assorted salad leaves

150g (5oz) good-quality tinned or bottled tuna
15 black olives with *herbes de Provence*
6 quail's eggs, boiled for 1 minute, then left to cool in the water, peeled and halved
6 salted anchovies, rinsed
1 tablespoon salted capers, rinsed (see page 52)
4 semi-dried tomatoes in olive oil, or fresh if unavailable, sliced

- Place all the ingredients for the dressing in a small plastic container with a tight-fitting lid and shake thoroughly until smooth.
- Plunge the beans into boiling water and cook for 5 minutes, then immediately cool in ice-cold water and reserve.
- Use the salad leaves as a base and then scatter the rest of the ingredients on top in an artful manner.
- Dress the salad immediately before serving.
- Serve with a baguette and a rosé wine from Provence.

SALTED CAPERS

As capers can equally well be pickled or salted, it was a toss up for us to decide in which chapter to deal with them. In the end we plumped for this one because the salted kind are widely considered to be superior as they maintain their taste and texture better. Either way, pizzas wouldn't be the same without them and tartare sauce wouldn't exist at all.

Capers are the unopened flower buds of the bush *Capparis spinosa*. Although it resembles a rose bush, it is actually closely related to the cabbage family. All over the Mediterranean, the buds are hand picked in the early mornings between May and July before they have a chance to open. They are then wilted in the sun for a couple of days before being put into bottles, either packed with salt or immersed in a strong wine vinegar brine. This gives them their tart piquancy. Raw capers don't taste of much at all.

Capparis spinosa is reasonably hardy, and can be grown outside as far north as southern Britain. If you like capers enough, this is certainly worth trying. If you do grow your own, you can copy the methods of the islanders of Pantelleria, south of Sicily, whose capers are probably the finest in the world (see page 47).

SICILIAN SALTED CAPERS

After harvesting, cover the buds with sea salt without bothering with a drying phase. Leave them for a week or so, during which time the liquid they contain will dissolve some of the salt to form a pickling brine. Then remove and drain them and salt them all over again.

Repeat the process twice, then pack them into sterilised jars in a ratio of 85 per cent capers to 15 per cent sea salt. You will need to rinse them thoroughly before using them or they may taste unbearably salty. To do this, place them in a sieve and hold it for a while under cold running water.

So long as the capers are packed in salt, they will keep for as long as 2 years.

NASTURTIUM PODS

If cultivating caper bushes sounds rather daunting, the good news is that you can make a decent poor man's version from the seed pods of the ubiquitous nasturtium. After the blossoms fall off, pick the green, half-ripened pods and drop them in the following mixture, which you will have boiled then cooled beforehand:

- 1.2 litres (2 pints) white wine vinegar
- 2 teaspoons salt
- 1 medium onion, thinly sliced
- ½ lemon, thinly sliced
- 1 teaspoon pickling spice
- 1 clove of garlic, crushed
- 4–6 peppercorns
- ½ teaspoon celery seeds

Your nasturtium blossoms are unlikely to all fall off at once, but you can keep adding the seed pods as they appear. Bottle them with the pickling liquid, keep them in the fridge and use them as you would ordinary capers. As they are immersed in liquid, they will not keep for as long as capers packed in pure salt. Use within a year.

GRAVADLAX

Gravadlax, gravlax or for that matter gravlaks all literally mean 'salmon from the grave'. Nowadays this Nordic delicacy is rarely preserved by actual burial. Instead, it spends a few days in a marinade of salt, dill and sugar before making its way onto the tables of many an upmarket restaurant. Traditionally carved in thicker slices than smoked salmon, gravadlax is a refreshing alternative to its exalted cousin and much easier to make. Excellent variations can be made with large sea bass and sea- or rainbow trout.

TO MAKE GRAVADLAX

Trim the salmon fillet and remove the rib and pin bones. The pin bones are the lateral ones which stick out at right angles to the skin along the fleshier parts of the fillet. If you can't see them, run your finger along the flesh and you will feel them. Use tweezers or pincers to remove them.

Place the salmon in a shallow dish, skin side down. Cover it with the curing ingredients and rub them thoroughly in. This isn't exactly traditional, but cover the dish with a layer of cling-film. Place a heavy chopping board or a smaller board weighed down with a pile of plates on the surface of the clingfilm and put it all in the fridge. Compressing the salmon helps to create a firmer, drier gravadlax.

After 24 hours, pour off the excess liquid and turn the salmon over. Return it to the fridge, weigh it down again and leave for another 24 hours.

The gravadlax is now ready. Uncarved, it will keep for another 5 days. To serve it, take a long, sharp knife and slice the fish diagonally, starting at the tail end. Greasing the knife with a smidgen of oil will make the carving easier. Gravadlax is delicious served with hot blinis and crème fraîche mixed with mustard.

1 salmon fillet, skin on
For every 1 kg (2¼lb)
  salmon you need:
50g (2oz) coarse sea salt
50g (2oz) granulated sugar
½ teaspoon freshly milled
  black pepper
zest of 2 lemons
50g (2oz) roughly chopped
  fresh dill

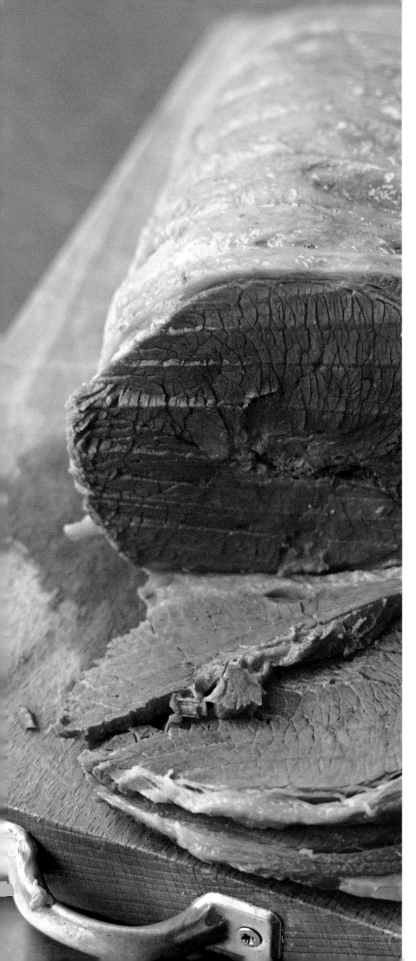

BEEF

The nomenclature of salted beef products is a food editor's nightmare. When Americans refer to 'corned beef', they usually mean what the British would describe as 'salt beef', namely whole cuts of cured meat. Meanwhile, when the Brits talk about 'corned beef', they mean highly processed South American beef sold in cans. To clarify, we use 'salt beef' to refer to the kind of thing you might find in a Jewish delicatessen, while 'corned beef' is what the Irish, and still more the Irish Americans, traditionally eat with cabbage on St Patrick's day.

OLD-FASHIONED CORNED BEEF

The 'corn' in corned beef has nothing to do with maize or wheat, but instead refers to the nuggets or 'corns' of salt traditionally used to preserve it. Nowadays, it is usually made with brine rather than dry salt, but the old name lives on. Once cooked, corned beef is fantastically tender.

- 3 cups sea salt
- 3.3 litres (5¾ pints) fresh water
- 4 cloves of garlic
- 1 large onion, rough chopped
- 2 tablespoons whole mustard seeds
- 2 tablespoons whole coriander seeds
- 1 tablespoon whole cloves
- 3 tablespoons whole peppercorns
- 2 large bay leaves
- 1 tablespoon thyme
- 1 beef brisket, about 2¼–3½kg (5-8lb)

You need to use a non-metallic vessel for brining. An enamelled churn would be ideal, as would a plastic or polystyrene ice chest.

Mix the sea salt and the water and stir until all the salt is dissolved. The brine is the correct strength when a fresh uncooked egg will float in it. If it doesn't, add ¼ cup of salt at a time until it does.

When the brine passes the egg test, add the remaining ingredients and the brisket. The meat needs to be submerged at all times, so weigh it down with a plate with a large stone on top.

Cover the brining vessel and refrigerate for about 10 days, turning the brisket over every other day. The thicker it is is, the longer it will take to cure. When the beef has finished curing, you can use some of it immediately and refrigerate (for up to 2 weeks) or freeze the rest for another day (it will keep for 6 months).

To cook the corned beef on its own, cover it with water, bring it to the boil, skim the surface and simmer for about 4 or 5 hours. It is ready when tender.

To make a traditional Paddy's Day dinner, take about 2kg (4½lb) of corned beef, place it in a receptacle suitable for long, slow cooking and cover it with water. Add 1 tablespoon of chopped fresh parsley, 1 teaspoon of powdered mustard, a couple of bay leaves, 4 cloves and about 10 black peppercorns. Bring this to the boil, cover the pot, turn down the heat and simmer for 3 hours. At the end of this period, skim the fat from the top and pour away half the water. Then add a large head of green cabbage, cut into wedges, and a dozen small, peeled onions. Place the beef on top of these vegetables and simmer on for another half an hour.

SALT BEEF

Although Europeans and Americans have been salting beef for donkey's years, Jewish chefs are the ones who have really got the recipe down pat. This is probably because salt pork, being non-kosher, was never much of an option to members of that faith. They therefore poured all their creative energies into mastering the preserving of beef.

The most interesting ingredient in the recipe that follows is saltpetre, also known as potassium nitrate. Saltpetre has played an important role in military history, partly through being liberally added to soldiers' food to curb their libido, partly as a major constituent of gunpowder. But in this context it has a much more benign effect. As well as allowing the meat to retain its bright pink colour when cooked and adding a distinctive and desirable flavour, it also inhibits the growth of harmful bacteria.

Getting hold of saltpetre can be problematic, as chemists are reluctant to sell it owing to its potential use in the manufacture of explosives. Any Jewish cook worth his salt will have a steady supply, but if you live in the UK and don't have such a contact, you can obtain saltpetre from The Natural Casing Company in Farnham, Surrey, telephone 01252 713545. Elsewhere, you may be better off looking for sodium nitrite (with two 'i's). This is just as effective and can be obtained from some licensed pharmacies. If you draw a complete blank, don't worry: the function of these exotic additives is largely cosmetic and you can make excellent salt beef without them.

INGREDIENTS
   1 heaped tablespoon salt
   1 heaped tablespoon saltpetre (if obtainable)
   2kg (4½lb) rolled beef brisket (Nick uses Scottish beef)
   2 litres (3½ pints) beef stock or water
   1 teaspoon black peppercorns
   4 bay leaves
   1 teaspoon salt
   ½ teaspoon caraway seeds
   2 carrots
   1 onion, cut in half

Dissolve the salt and saltpetre in a cup of warm water.

Place the meat in a bowl just big enough to contain it and immerse in cold water, adding the dissolved salt and saltpetre.

Cover the bowl and leave it in a cool place for 3 days.

After this period, the water will be tinged pink and the meat itself will be a dull red. Remove the beef and rinse it well.

Simmer the beef in beef stock or water along with the peppercorns, bay leaves, salt, caraway seeds, carrots and onion for around 3 hours until the beef is nice and tender. (You can reuse the stock to make sauce or a classy borscht.)

Let the beef cool in the stock to keep it juicy, then remove it and store it in the fridge for up to a week, or in the freezer for up to 6 months. Now you can make some sublime sandwiches.

SALT-BEEF SANDWICHES WITH DILL PICKLE

The salt-beef sandwiches beloved of New Yorkers and anyone else with access to a good Jewish deli are traditionally made with dill pickles. The green cucumber complements the pink meat and livens it up considerably. If only Jews were allowed to eat salt-beef sandwiches, there would probably be a mass conversion. **Makes 4 sandwiches**

THE DILL PICKLE

6 large pickling cucumbers, sliced and de-seeded

2 sprigs of dill

4 cardamom pods

6 cloves

12 peppercorns

x2 800ml (28fl oz) jars

1 large loaf of rye bread, preferably with caraway seeds, toasted

Many thick slices of salt beef

Dijon mustard, mixed 50/50 with mayonnaise

- You can either buy in some dill pickles or make your own. Dill pickle is made much as the pickled cucumbers on page 115, only the tarragon (which is what Nick happened to have around when we were doing the 'Pickling' chapter) is replaced with dill and the bay leaves and caraway seeds are left out. Make a pickling brine with water, salt and cider vinegar as on page 115. Then pack your sliced and de-seeded cucumbers into the jars. Add a sprig of dill, 2 cardomom pods, 3 cloves and 12 peppercorns. Pour in the pickling brine, filling the jars to their necks, then seal and leave in the fridge for a good month before use.
- Take 4 slices of toast and spread them with the mustard mayonnaise. Make a pile of salt beef on each, crowned with a not-insignificant mound of dill pickle. Place the number two slices of toast on top, cut each sandwich in half and serve.

SWEET-CURED HAM

We can't call the following recipe 'Parma Ham' because that would be pretentious and land us in trouble with the EC. However, this worthy approximation has much in common with the well-known Italian delicacy and its noble Spanish cousin *jamon serrano*. This will become obvious when you carve it into melting, wafer-thin slices and serve it with melon or fresh figs. Nothing has given us more delight than making this at home.

Although you will rightly be infinitely proud of your finished ham, you don't want to go to all this trouble with meat from anything other than a fit, happy, organically raised pig. This will be fattier than a ham from the equivalent intensively farmed animal, but it will be all the more tasty as a consequence.

1.8–2.7kg (4–6lb) coarse salt

900g (2lb) sugar

1 teaspoon saltpetre, if available (only really for the colour)

1 fresh ham, skin still on – the one currently hanging in Nick's shed weighed 9kg (20lb) when purchased

½ bottle red wine

100ml (3½fl oz) balsamic vinegar

¼ cup fresh pork fat

1 large head of garlic, peeled and blended or finely chopped

1 tablespoon freshly ground black pepper

1 tablespoon additional salt, for the rub

Some muslin

A large paper bag

Mix the salt, sugar and saltpetre to make a sweet-cure. Rub well into the ham, particularly around the exposed bone.

Place the ham in a wooden box and store it in a cool place (we use a shed) for 5 days for each kilo (2¼lbs) of meat. Check it every couple of days to make sure it remains well covered with the curing mix. If necessary, repack this around it.

At the end of the curing period, remove the ham from the box, rinse thoroughly with water and dry with a cloth.

Combine the wine and balsamic vinegar and boil until reduced by half. Use this liquid to give the ham a good washing. Tie a strong cord to the shank and leave it in a well-ventilated place to hang naked for 24 hours.

Mix the pork fat, garlic, pepper and the 'new' salt together and rub the resulting paste all over the ham.

Now wrap the ham in muslin and place it in your paper bag, which should form a loose shroud. Hang in a cool, airy place for 3 to 6 months, or longer if you can bear the wait.

The flavour will intensify over time. When you fancy a slice, trim away the exterior of the portion you are about to carve, and use a long, sharp knife. You can then cover the ham and hang it up again. As long as it remains dry, it will keep for 6 months.

SMOKING

Smoked foods are delicious. Period. Impregnating a fish, fowl or hunk

of meat with fragrant hardwood smoke concentrates the flavour and

transforms the colour and texture. In fact, your authors are hard

pressed to think of any of the regularly smoked foods that isn't

improved by the process.

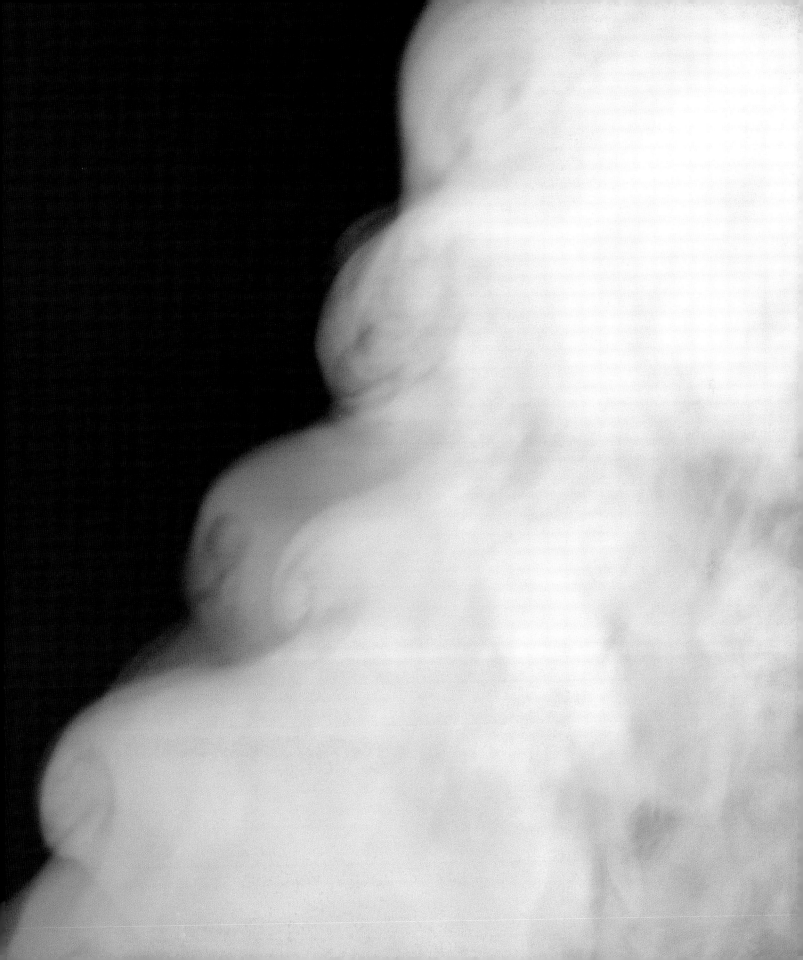

Like most of the techniques in this book, smoking was discovered by accident. At some point in the distant past (and archaeology suggests people have been using fire for at least 400,000 years), someone discovered that pieces of meat left hanging in whatever smoke-filled cave or hovel they were living in at the time remained edible for much longer than might have been expected. They tasted great too. Before long, it also became apparent that smoking food could save lives during times of scarcity.

By the Middle Ages, many northern European farmhouses had special smoking shelves built into their chimneys and in some coastal areas smoking was already established as an industrial process.

THE SCIENTIFIC BIT

The preservative powers of smoke rest on a number of exotic constituents, including antioxidants like butylated hydroxyanisole and various alcohols and phenols. Although the quantities involved are minute, these are not all the healthiest of substances, so it would be unwise to subsist entirely on smoked foods.

As a kind of double insurance policy, most smoked foods are dry salted or brined for a period prior to the main event. This curing reduces their water content, firms up the flesh and seasons it. Nowadays, with the advent of freezers and vacuum packs, flavour is relatively more important and longevity less so than they once were, so smoked foods are typically salted and smoked for shorter periods.

HOT AND COLD SMOKING

The crucial division in home-smoking is between cold and hot. In the former, the temperature in the chamber is not allowed to rise much above that of an ordinary room, drying the food but not cooking it. Typical cold-smoked products include salmon and fillet of beef. If the temperature rises above 29°C/85°F while you are smoking them, you should start to worry. Fish, in particular, may start to disintegrate, and microbes to proliferate. Hot smoking, on the other hand, partially cooks the relevant items and usually requires a temperature of between 82°C/180°F and 93°C/200°F for fish (such as mackerel), and a rather broader range of 82°C/180°F) to 115°C/240°F for poultry and meat. Beware of the grey zone between these 'bandwidths'. We'll deal with hot smoking later in the book, but for now we'll concentrate on cold (incidentally, most hot-smoked foods require a preliminary period of cold smoking before stage two).

COLD SMOKING

Cold-smoking doesn't actually cook food, it just flavours it and gives it a long shelf-life. Some cold-smoked products, like smoked salmon and beef, are best eaten raw. Others, like kippers, you'll want to cook before consumption.

THE SIX STAGES

1) Weighing. The best way to tell whether an item is sufficiently smoked is to weigh it and compare the figure with its initial weight (that is, after removing surplus fat but before brining or salting). You should make a note of the original weight or you may forget it. By and large, fish need to lose less weight than cuts of meat, but we'll give you the appropriate percentages for each product with the relevant instructions.

2) Brining or Dry Salting. This is vital as it removes moisture from the food and renders it unappetising to bacteria. Brining is sufficient for most smoked products – the only item we've instructed you to dry salt is bacon, and even this is negotiable.

3) Drying. Whether you've dry salted or brined your victuals, you'll need to hang them for a period prior to smoking them or they will be too moist to smoke properly.

4) Smoking. The main event. As mentioned in the spiel at the start of this chapter, you need to make sure the temperature inside your smoker never rises above 29°C/85°F during the smoking phase.

5) Maturing. After you've finished smoking, the flavourants from the smoke will continue to penetrate the flesh for a while. Most smoked products will benefit from a good 24 hours of maturation before consumption.

6) Storage. Smoked foods will keep in the fridge for at least a week and quite often a good deal longer, but they also freeze well. Wrap them in foil before stashing them in the freezer.

The instructions for smoked salmon, the first cold-smoked delicacy you will come to as you read on, provide a detailed example of the standard sequence in action.

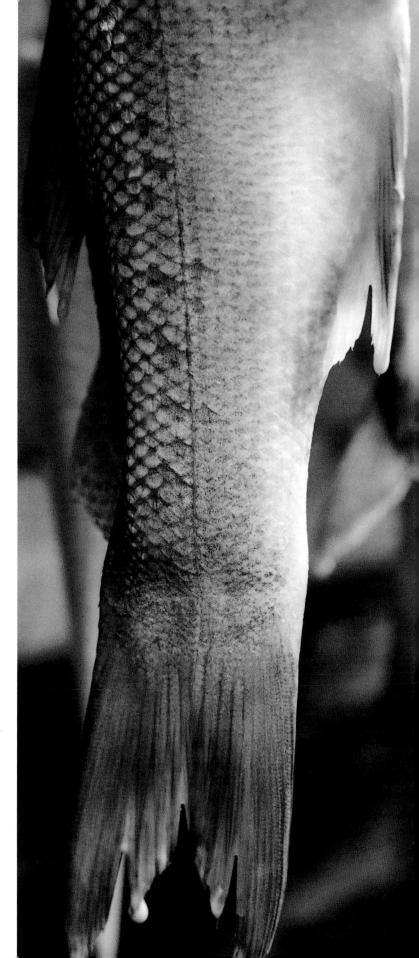

BUILDING A COLD SMOKER

In the simplest terms, what you need is a sufficiently large chamber to allow you to hang or support a decent-sized fish from somewhere near the top with room to spare, a ventilation hole or holes near the top to act as a chimney, a hearth or metal plate to house the sawdust, and a perforated piece of wood or metal a few inches above to disperse the smoke evenly as it rises. Once you grasp the principles, there are many ways to skin the cat. Here we'll teach you to build a simple wooden or breezeblock smokehouse from scratch. This is a most satisfying undertaking.

THE TWO ALTERNATIVES

The first consideration is the method of delivering the smoke. The two alternatives are lighting a smouldering fire inside the smoker itself or piping it in from a remote chamber, à la Nick's glorious pictured device. The first approach is only advisable if the smoker is rather large: you don't want the burning material to significantly increase the internal temperature. If you choose the latter method, go to the local DIY shop and buy a length of flexible aluminium tubing 10 to 15cm (4 to 6in) in diameter to channel the smoke in the desired direction.

THE SMOKING CHAMBER

Turning to the main chamber, an ideal size for domestic use would be in the order of 1.5m (5ft) high by 90cm (3ft) broad and deep. If you use breezeblocks, the construction is child's play. If you prefer to use wood, which is more attractive, but obviously more combustible, make sure it hasn't been treated with anything vile that might taint the food. Buy enough planks to do three sides plus a roof if you'll be building a door, or four sides if you're more inclined towards a lid (see photograph). The chamber will be significantly sturdier if you reinforce each side by nailing three beams to the planks in a 'Z' shape.

BUILDING THE FRAME

The first step is to build a frame. You will need 4 x 90cm (3ft) pieces of wood for the base, 4 x 90cm (3ft) for the top and 4 x 1.5m (5ft) for the corner posts. The next stage is to nail the planks to the frame, the exact configuration being dependent on

your preferred method of access. If you are handy at this sort of thing, you might want to consider installing a door, otherwise a wooden lid (preferably hinged) is perfectly adequate for getting your foodstuffs in and out.

Next, install two battens on the inside of the box, about 45cm (18in) from the ground, to support the smoke diffuser. If you intend to produce your smoke inside the chamber, you'll need another two below them to house the metal plate on which your wood will smoulder. Leave enough room underneath for ignition purposes (see next paragraph). If you are planning to rest your fish and meat on racks, you will need a couple of pairs of extra battens higher up the chamber. If, on the other hand, you plan to hang them, you can either drill small holes in the sides and pass appropriate metal hanging rails between them, or, following our example, screw some hooks into your hinged lid. If you do this, you want to make sure you can't or don't open it a full 90 or 180 degrees, or the food you are smoking will bang against the lid and any unoccupied hooks.

PRODUCING THE SMOKE

The final piece of the jigsaw is to sort out your smoke source. You will need to make some vent holes a couple of inches below the roof to draw the smoke up and out. If you're going for the remote fire-pit technique, you'll need to cut a hole that is the diameter of your aluminium tubing on one side, near the bottom, and secure it. Otherwise, you'll want to ignite your fuel, probably in the form of sawdust, by placing a small gas burner underneath the metal plate on which it is to rest. If you leave a gap at the

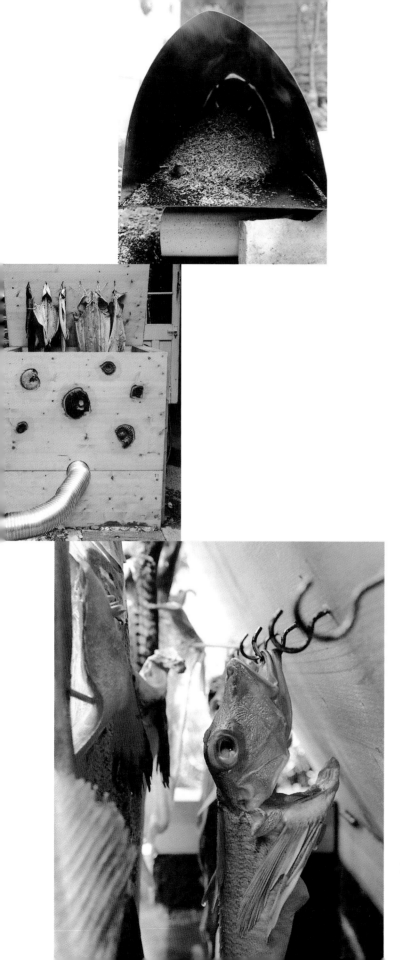

bottom of the front of the smoker (perhaps by using slightly shorter planks on this side and nailing them to an extra crossbeam a few inches above the bottom), it will make this considerably easier.

Don't be daunted. It's easier than it sounds, and if your smoker leaks in a few unexpected places, it's no big deal. In fact, the hardest part of the process in our experience is finding a reliable source of fuel. By far the most user-friendly material is sawdust. We'd recommend that you purchase a cheap electric planer to make your own. A couple of biscuit-tinfuls will smoulder happily for a good 10 hours. The crucial thing is to avoid softwoods such as pine, which will give your food an unpleasant, antiseptic taste. We tend to use oak, but beech, birch, hickory or any fruit wood will do just as well – perhaps better, as they tend to produce more smoke and each imparts a subtle flavour of its own. You should be able to buy hardwood planks from timber merchants.

SAFETY PRECAUTIONS

Igniting your sawdust can be tricky – you don't want it to burst into flame – but you'll soon get the hang of it. Some strategic blowing may come in handy. As mentioned previously, smoke produced in either of the suggested ways is unlikely to heat your chamber to danger level, but you might want to purchase a jam thermometer just to make sure. Above all, refrain from smoking in hot weather. If the outside temperature is above 26°C/80°F, you don't have much of a chance. It is no accident that the traditional time of year for smoking is the autumn.

SMOKED SALMON

If you are anything like us, the first thing you'll be itching to smoke is a salmon. The undisputed king of smoked foods, S.S. is often pretty good if you buy it from the shops. If you make your own, it can be sublime. You'll also be very popular with your friends. There is one more consideration. Shop-bought smoked salmon tends to be four or five times the cost of the fresh equivalent. Why is anyone's guess – making it is virtually free. 'Nuff said.

It would be faintly sacrilegious to smoke a wild salmon unless you found yourself in the Arctic with a mountain of them, as in Britain at least there are hardly any left. But you do want to buy the freshest, firmest farmed fish you can lay your hands on.

SPLITTING/FILLETING

You have a choice between filleting the fish or splitting it. Filleting is the better option in the case of very large salmon or if you plan to lay the fish on a rack during smoking. Your fishmonger will happily do it for you. To split a fish, take a sharp knife and cut the animal down the back from head to tail. Remove the guts, gills and blood channels, particularly the large one running along the spine, and wash the fish thoroughly.

WEIGHING

Once you have gutted the fish, weigh it. This will allow you to determine when it is done. The salmon should diminish in weight by 17–18 per cent during the smoking process. In other words, a fish which tips the scales at 5kg (11lb) after gutting should weigh around 4.125kg (9lb 1oz) once smoked.

BRINING

An 80 per cent salt solution is about right for this stage. This equates to 1.2kg (2lb 10oz) of salt to every 4.54 litres (1 gallon) of water. Make sure you use sea salt. You will need to weigh down the fish to prevent it floating to the surface. Leave the fish in the brine for 1 to 3 hours, depending on its size and fat content (fat salmon will need a longer immersion time than skinny ones).

DRYING

At this stage, you need some string. Use it to suspend the salmon via the tail, thread it through the eye sockets, or pass the string through incisions made under the shoulder plates. You will also need wooden skewers or small hazel twigs to hold the fish open. These can be easily inserted into the flesh. Hang the fish to drip-dry in the inactive smoker for 24 hours. Don't rinse off the brine.

SMOKING

The desirable duration of smoking depends on three things: a) personal taste, b) ambient temperature and c) humidity. The colder or more humid the weather, the longer the process. On average, you are looking at anything from 24 hours to 2½ days. The acid test, as mentioned above, is to weigh the fish from time to time, aiming for a 17–18 per cent reduction.

MATURING

You should refrain from eating the salmon for a good 24 hours, to allow the smoky surface deposits to work their way inside.

STORAGE

If you keep it in the fridge, your smoked salmon will be in tip-top condition for 5 days and more than edible for 10. It also freezes well, provided you first wrap it in aluminium foil.

SMOKED SALMON, NOODLES AND SWEET CHILLI SAUCE

This oriental-inspired dish is an excellent way to use some of your proudly smoked salmon. The mild, sweet chilli sauce is slightly gelled and tinged pink. It perfectly complements the amber translucence of the salmon, as do the glassy noodles, and it keeps in the fridge for months. **Serves 2**

125g (4½oz) rice noodles

110g (4oz) baby greens or shredded spinach

Salt

2 teaspoons sesame seeds

2 heaped tablespoons sweet chilli sauce (see page 177)

100g (3½oz) smoked salmon, cut into strips

- Make the sweet chilli sauce following the recipe on page 177.
- Cook the rice noodles as per the instructions on the packet, then chill in cold water and drain thoroughly.
- Plunge the greens/spinach in boiling salted water for about a minute, then chill in iced water and drain as above.
- Dry-fry the sesame seeds in a small pan over medium heat until golden (approximately 2 minutes).
- Throw the noodles, sweet chilli sauce, salmon and baby greens together in a large bowl and mix. If desired, you can heat the salad slightly in the microwave.
- Divide between two plates, scatter the sesame seeds on top and serve.

KIPPERS

The Kipper, served with toast and marmalade, is such a quintessential British breakfast item that you'd have thought it had been around for ever. But amazingly, the kippered herring was only invented in 1843. Herrings had long been super-abundant in the North Sea – in many places you only had to dip your hand in the water to catch one – but hitherto the standard method of preserving them had been to hot-smoke them whole and ungutted and long and hard. The result was the infamous red herring. It was the appearance of the kipper that consigned it to its proverbial elusiveness.

The word 'kipper' originally referred to a tired old salmon that had spawned. The Scottish just tended to split such beasts and smoke them overnight, and from the fifteenth century onwards they had described the practice as 'kippering'. But it was a Northumberland fish merchant named John Woodger who, in his search for a delicate, less salty alternative to the red variety, first thought of applying the technique to the herring. Splitting the fish allowed a much shorter period of salting, while improved transport links and the advent of refrigeration removed the need to smoke them to oblivion. The modern kipper had arrived.

SPLITTING AND WEIGHING
To make kippers at home, get hold of the freshest herrings you can find and split them and splay them. This is done by gutting them, including the gills, then cutting through the connecting rib bones all the way along one side of the backbone. Place the fish belly-down on a clean surface, then press their backs firmly with your hand. This will split them. Then you should weigh them and make a note of your findings.

BRINING
Next, make around 3 litres (5¼pints) of brine using 220g salt per litre of water (or 4½oz per pint) and immerse your fish in this for 15–20 minutes depending on how salty you like them.

DRYING
The next step is hanging the fish to dry. At this stage, you have a choice to make. If you plan to dry and smoke them on a rack, their own weight will keep them splayed during the process, provided you lay them like an opened book on a photocopier. If, on the other hand, you mean to hang them, gently insert some soaked wooden cocktail sticks in their flesh to keep them open. You can hook them into hanging position by their mouths. Either way, you want to leave them to drip-dry for a couple of hours in the smoker (but without the smoke).

SMOKING
Smoking the kippers should take between 8 and 16 hours. The desired weight loss is 14–18 per cent. The optimum smoking temperature is 26°C/80°F. You should make sure it never rises more than a few degrees above this, except for the last half hour, when you can ramp the temperature up to 35°C/95°F or slightly below to give the fish a nice glossy finish.

STORING AND EATING
Your kippers will be fine in the fridge for 4 days to a week. They are best grilled, but you can also fry them or bake them. Nick makes a tasty salad consisting of kipper chunks, baby broad beans, new potatoes and hard-boiled egg doused in a honey mustard dressing. Watch out for bones. The British spent many unnecessary years worrying that the Queen Mother would be finished off in this manner.

SMOKED COD'S ROE

When Nick confessed to having hot-smoked his cod's roe not so long ago, Johnny refused to talk to him for weeks. Nick stubbornly maintains that it was delicious on rye bread with a little sour cream, but now confesses that cold-smoking is the 'proper' way to go.

You can use either fresh or frozen roes for smoking purposes. Cod's roes are extremely delicate, so you need to be gentle with them. If you convert the finished articles into taramasalata, you can ponder the mystery of how it became a Greek national dish despite the absence of cod in the Mediterranean.

Get hold of some pristine, unblemished cod's roe. An average specimen weighs about 250g (9oz). Handle the roe with care or they may rupture and cover you in eggs.

BRINING
Wash the roe, then soak them for 1¼ hours in a 70 per cent brine, which equates to 220g salt per litre of water (4½oz per pint).

DRYING
Drying the roe can be tricky as they are so fragile, but provided you are careful you should be able to thread some string between the sacs (which come in pairs) and use it to hang them from hooks. Alternatively, you can lay them on racks, but the wires will leave conspicuous indents. Drip-dry the roe for a couple of hours in the 'inactive' smoker.

SMOKING
The roe should then be cold smoked at 26–29°C/80–85°F for 24 hours. The end product will have lost about 25 per cent of its original weight and be firm, dry and easy to slice. You can store it in the fridge for up to 10 days or wrap it in foil and freeze it for up to 6 months.

TARAMASALATA
This is one of the more rewarding ways to use your smoked cod's roe.

- 100g (3½oz) smoked cod's roe, membrane removed
- 2 slices of white bread, crust removed
- 150ml (¼ pint) olive oil
- Juice of 1 lemon
- A little warm water
- Salt and pepper

Blend the cod's roe and bread in a food-processor until smooth, then remove and place in a bowl.

Gently pour in the olive oil and lemon juice, whisking constantly with a hand or electric whisk. After they have been added you may want to add a little water for a looser consistency.

Season with salt and pepper to taste.

SMOKED BEEF

Smoked beef, like bresaola, is eaten raw, so cold-smoking is the appropriate method to use for it. Fillet is the best cut, preferably the thicker part which used to be further from the animal's tail. You don't need it to be top quality: you can buy meat taken from older-than-usual cattle if you have a friendly local butcher. This will be cheaper than prime fillet, though of course it needs to be in good condition for what it is.

WEIGHING

Your fillet should have been hung for at least a week by the butcher to make the end-product suitably flavoursome. Remove most of the fat or it will hinder smoke penetration and may turn rancid. Then weigh the fillet. It should lose 20–25 per cent of its weight during the smoking process. Prick it deeply in several places to ensure that the brine in which you are about to immerse it penetrates it properly.

BRINING AND DRYING

To make the correct strength of brine, dissolve salt in water at the rate of 1.2kg per 4.54 litres or 2lb 10oz per gallon. The salt will dissolve better if you boil up about a third of the water and stir it in before adding the rest of the water cold.

Cover the fillet with brine and leave it to soak for about 3 hours (a little more if it is thick, a little less if it's on the thin and lean side). Remove the meat from the brine and tie a loop of string tightly round it a few centimetres/couple of inches from one end. Use this to hang it to drip-dry for 24 hours before smoking. As usual, your smoker is a good place to do this.

SMOKING

The smoking process will probably take 5 to 8 days depending on the temperature. Make sure that this does not exceed 26°C/80°F or the fillet will harden, preventing the necessary moisture loss. After 4 days or so, start weighing the beef from time to time to see how things are progressing. It will be thoroughly black on the outside by this stage. When the fillet has lost the requisite 20–25 per cent in weight, the final test of its readiness is to cut it through the thickest part and inspect its exposed surfaces. The meat should be matt rather than shiny. If it isn't, give it another day or two in the smoker.

STORING AND SERVING

Your smoked beef will keep in the fridge for 3 or more weeks, but you can always wrap it in foil and freeze it. This way, it will keep for 6 months. To serve it, slice it thinly across the grain. Like smoked salmon, it is very good with lemon juice and brown bread.

SMOKED BEEF WITH CREAMY HORSERADISH

Horseradish is well known for its partnership with roast beef, but it goes equally well with the smoked kind. This is a luxurious version of the sauce, fiery yet velvety. You'll end up with more than you need, so consider trying it with smoked mackerel (page 78). The beef should be sliced tissue-thin. **Serves 2**

Lots of thin slices of smoked beef fillet (see left)

A pile of wild rocket

Plump ripe cherry tomatoes

A small loaf of country bread and good creamy butter

Lemon juice for the beef, if desired

THE SAUCE

300ml (½ pint) chicken stock

150g (5oz) peeled horseradish, freshly grated (make sure you wear a mask)

300ml (11fl oz) double cream

50g (2oz) fresh white bread, cut into cubes

10g (½oz) mustard powder

½ teaspoon salt

¼ teaspoon ground white pepper

1 egg yolk

- To make the sauce, heat the chicken stock in a saucepan along with the horseradish. Bring to the boil and simmer for 15 minutes until the chicken stock has lost about a third of its volume.
- Add the cream and simmer for about 20 minutes more, stirring frequently.
- Add the bread slowly, stirring constantly, until it has all been smoothly incorporated. This should take about 5–10 minutes.
- Mix the mustard powder, salt and pepper with 1 tablespoon of water and stir until the lumps have disappeared. Then whisk it into the hot sauce.
- Finally, remove the sauce from the heat and stir in the egg yolk.
- Pour the sauce into sterilised pots with tight-fitting lids and cool in a sinkful of iced water. The quicker you do this, the longer it will keep – for 2 weeks in the fridge if you've done it properly. Once a pot has been opened, however, the horseradish must be eaten within 3 days.
- Get all the ingredients together on a plate as soon as possible. Find a long, balmy summer's evening and enjoy with a pint of ale.

SMOKED BACON

This recipe is for pancetta-style bacon made from pork belly. Decent bacon is hard to come by in the supermarkets, which tend to pump it full of water to add bulk. This produces tell-tale white goo in your frying pan. Organic bacon is increasingly available, but for some reason it is rarely both smoked and streaky. This is difficult for us as we're no great fans of back bacon. 'Too little fat!' we cry as we tuck into our plates of lard. Generous seams of fat not only add flavour, they also render down to make an excellent frying medium. For this reason alone, you are better off with pork belly from free-range pigs, which tend to be fatter than their unfortunate fellows from factory farms.

We wouldn't attempt this with less than 10kg (22lb) of pork belly, partly because it will lose up to 30 per cent of its pre-salted weight during smoking, partly because it is so good that you may end up proudly giving a lot of it away. A meat-filleting knife will prove invaluable.

TO MAKE PANCETTA

- 4kg (9lb) sea salt
- 4kg (9lb) sugar
- 10kg (22lb) pork belly, ribs and gristle removed (your butcher should be happy to remove the ribs and gristle or you can do it yourself with a filleting knife. If you do, buy slightly more pork belly to compensate)
- 1 large branch of dried rosemary
- 1 large tupperware/non-reactive container

Mix the salt and sugar together.

Sprinkle a layer of salt mix at the base of the container, then a layer of pork, then another layer of salt and so on, finishing with a layer of salt. Continue until all the pork is used up and packed hard with the sweet-cure.

Leave in a cool place to cure. After 2 days, pour away any excess liquid and re-pack the pork upside down and in reverse order (i.e. the seam of meat that was highest should now be lowest). If too much of the sweet-cure has dissolved, you can top it up with some more.

After 3 more days wash the pork off with cold running water, then sprinkle with a little sea salt and rosemary.

Air-dry in a cool area for 24–48 hours.

Cold smoke for 48 hours at 24–26°C/75-80°F. We pierce the meat and thread string through to hang it from hooks, but you can use racks if you prefer. Wood-wise, oak is great, but cherry, beech or sweet chestnut are also good.

Your bacon is ready for frying immediately. Nevertheless, we like to leave it to mature in the fridge for a month, by when it is a deep, almost translucent amber.

The bacon can be stored in the fridge for up to 3 months, but make sure you keep it dry. Look out for dark moulds. If any form, throw the meat away.

Before using this pancetta-style bacon, trim off the skin. You can either slice it very thinly to make breakfast rashers or dice it for use as an ingredient in sauces and stews.

PEA AND SMOKED PANCETTA SOUP

Pulses and bacon products have a natural affinity. Here, the salty tang of pancetta is offset by a naturally sweet petit pois purée. Serves 4

150g (5oz) diced smoked pancetta (see page 71)
75g (3oz) leek, washed and roughly chopped
150g (5oz) onions, roughly chopped
25g (1oz) butter
1.2 litres (2 pints) chicken stock
400g (14oz) potatoes, peeled and diced

1 teaspoon salt or more according to taste
¼ teaspoon ground white pepper
¼ teaspoon ground bay leaf or 2 whole ones
500g (1lb 2oz) petit pois
A sprig of mint, chopped
150ml (¼ pint) double cream

- Fry the smoked pancetta over a moderate heat until nicely browned. Remove and reserve.
- In the same pan, sweat the leek and onions in the butter and the fat given off from the pancetta until soft.
- Pour in the chicken stock, then add the potatoes and simmer for 20 minutes.
- Add the salt, pepper, bay and petit pois and simmer for a further 10 minutes.
- Blend the soup with a hand blender until smooth. If you are using whole bay leaves, remove them before blending.
- Add the mint, cream and pancetta and serve immediately.

SMOKED EGGS

It may never have occurred to you to smoke eggs but now is the time to consider it. It is difficult to say what the process does to their flavour without using the word 'smoky', but it makes them a lot tastier and gives them a remarkable colour. If you smoke a few when you get up, they will be ready for a late tea.

Don't use newly laid eggs as they are difficult to peel. Quails' eggs are at least as good smoked as hens' eggs and need a lot less time in the smoker.

PREPARING THE EGGS

To smoke either kind of egg, place them in a pan of cold water, bring quickly to the boil, remove from the heat and leave them in the water to cool. With quails' eggs, you need to arrest the cooking process by refreshing them under cold water after 3 minutes.

Peel the eggs before smoking and season them with salt and white pepper and perhaps a little soy sauce. Place them on racks in your smoker and smoke them at about 26°C/80°F. Hens' eggs take about 12 hours, quails' eggs between 4 and 6. They will keep in the fridge for up to 2 weeks.

SMOKED HADDOCK AND EGG PIE

Smoked eggs are excellent in salads and sandwiches and go particularly well with smoked haddock in a pie. To make one to feed 4 people, boil and mash 6 medium potatoes with 100ml (3½fl oz) milk, 40g (1½oz) butter, salt, pepper and 125g (4½oz) grated mature cheddar. Dice 250g (9oz) smoked haddock and mix it with 6 diced smoked eggs, 350ml (12fl oz) béchamel sauce, some chopped parsley and a little grated nutmeg. Spread the potato on the fish mix and grate some cheddar on top. Bake in the oven for 45 minutes at 200°C/400°F/gas mark 6. This recipe cries out for peas as an accompaniment.

SMOKED CUSTARD TARTS

This isn't the most obvious recipe on the planet, but try it and you may be pleasantly surprised. Makes 2

THE PASTRY
500g (1lb 2oz) '00' grade plain flour
175g (6oz) icing sugar
250g (9oz) butter
1 medium egg, plus 2 egg yolks
½ teaspoon natural vanilla extract

Sift the flour and icing sugar into a large bowl. Work in the butter with your fingers until soft and crumbly. Then mix the egg, yolks and vanilla together, make a well in the flour, then stir in the egg until a soft dough is formed.

Loosely roll out the dough and divide it into 2 balls. Cover in clingfilm and leave to rest in the fridge for 30 minutes before use.

Take two 20cm (8in) flan tins with removable bases, about 4cm deep. Roll out the dough and press it out into the tins leaving it hanging slightly over the lip. Weigh it down by cutting out a circle of greaseproof paper, placing it on the uncooked pastry and evenly spreading some beans or pulses on top.

Bake for 10 minutes at 180°C/360°F/gas mark 4. Remove the beans, trim the pastry and cook on for a further 15 minutes.

THE CUSTARD
8 medium eggs
500ml (18fl oz) milk
500ml (18fl oz) double cream
250g (9oz) caster sugar
2 teaspoons natural vanilla extract
Freshly ground nutmeg

Lightly whisk the eggs, and then add the milk, double cream, sugar and vanilla extract and hand-beat for 30 seconds.

Fill the tart shells, then bake in the oven at 150°C/300°F/gas mark 2 for 45 minutes or until the custard has just set.

Cold-smoke the tarts for 2 hours and serve.

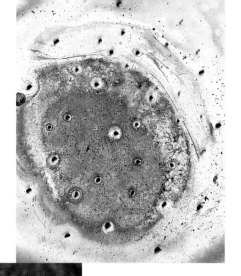

HOT SMOKING

Hot smoking, in contrast to cold, cooks whatever is subjected to it. As mentioned previously, hot-smoked products almost always undergo a period of cold smoking beforehand. It is during this phase that most of the moisture loss and the greater part of the smoke flavouring take place.

Hot-smoked fish need to be processed within quite a narrow temperature range, namely between 82–93°C/180–200°F. Any colder and they enter the danger zone for bacterial growth; any hotter and they start to disintegrate. Meats are more tolerant. They can be successfully hot smoked at anything between 82–115°C/180–240°F. Some products, particularly oily fish, are suitable for both kinds of smoking. It is worth experimenting to see which you prefer. Others can only be successfully cold smoked – cheese, for instance, will simply melt if placed in a hot smoker. Still others, including poultry and most kinds of game, are only really amenable to hot smoking. It is worth bearing in mind that hot smoking takes much less time than cold.

BUILDING A HOT SMOKER

The simplest way to build a hot smoker, and the one we would recommend, is to adapt a galvanised dustbin or dustbin-incinerator. All you need to do is punch several holes in the bottom of the bin, perhaps including a large central disc, and then drill parallel holes in the sides at three levels to support, going upwards, a dripping tray and two wire racks on which to place your food. These can rest on steel rods passed through the appropriate holes. As far as roofs go, if you are using an adapted regular dustbin, a sheet of wood will do fine. The smoke will escape through the holes in its side. If you are using a dustbin-incinerator, it will come with a ready-made chimney in its lid. This can be blocked off with a piece of wood if you are having trouble reaching the requisite temperature.

Next, you stand the bin on two large breezeblocks and place two smaller bricks just inside them to support a metal plate on which your sawdust will smoulder. You can easily make one by flattening a piece of corrugated iron. If this is perforated, it will make ignition easier. Finally, you place a gas-ring under the hearth plate to light the sawdust and maintain an appropriate temperature within the chamber during smoking. You may need to adjust the flame in the course of your smoking period. When the food is introduced into the chamber, the temperature is likely to drop, and once the sawdust gets going it will contribute heat of its own, so, other things being equal, you will want to turn down the gas a little. As with a cold smoker, a jam thermometer will allow you to monitor progress.

SOME SMOKING TIPS

Finally, some words of advice. Before you process any food in your chamber, pass smoke through it for at least 12 hours to coat it with tar deposits. This will prevent tainting. And, mindful of the fact that hot air rises, it is a good idea to swap your trays around half way through smoking to ensure their contents receive the same amount of heat. The good news is that the hot-smoking process is immune to the outside temperature.

HOT-SMOKED CHICKEN

The first time Nick hot smoked chicken, in his naivety he performed the operation with no brining and no initial period of cold smoking. He served the chicken hot to his guests after a 3-hour smoke at 110°C/230°F and they were delighted with it. He had essentially used the hot smoker as an oven in the procedure known as smoke-roasting. The chicken was tender and juicy and only mildly smoky as the smoke didn't have much time to penetrate, plus the skin quickly dried to form a barrier.

Traditionally hot-smoked chicken has an altogether more pronounced smoky flavour and is drier and firmer. It can be kept in the fridge for about a week or in the freezer for 6 months. It will revolutionise your Caesar salads.

BRINING AND DRYING

Take a medium-sized chicken and prick down to the bone with a fork to assist brine penetration.

Prepare an 80 per cent brine (265g salt per litre/5.4oz per pint of water) and immerse the bird in it for 2 hours.

Dry the chicken for 24 hours in a cool, airy place. Make a loop of string, insert it under the wings and hang it from a hook.

THE SMOKING STAGES

Cold smoke the chicken for 48 hours, making sure that the internal temperature of the cold smoker does not exceed 25°C/77°F. Then hot smoke the chicken for 2 hours at 110°C/230°F.

Hot-smoked chicken is well complemented by piccalilli (see page 118) and sour dough bread.

HOT-SMOKED DUCK AND GOOSE BREAST

Duck and geese may certainly be smoked whole, but we tend to limit ourselves to breasts for a variety of reasons. For one thing, you can fit more of them in the smoker. For another, there will be less molten fat to deal with. Nick found out the wisdom of this after a conflagration which consumed 3 sausages, a mackerel, a duck and an old pair of gardening gloves. Another advantage of sticking to the breasts is that the other parts of the creature can be put to better use, the legs and wings to make confit or rillettes (see page 204), the carcasses to be roasted as the basis for excellent stocks.

Both ducks and geese smoke very well. Hot-smoked 'fresh' for an hour at 110°C/230°F, they make a delicious starter (served warm) or classy ingredient for a cassoulet (see page 192). Alternatively they can be hot smoked in the traditional way: that is, after a period of cold-smoking. This takes longer but is definitely worth it.

The larger the breasts, the better the end results. First, prick them thoroughly and immerse them in an 80 per cent brine (see page 64) for 2 to 3 hours depending on their size.

Then dry them on a rack for between 12 and 24 hours in your inactive smoker. Cold-smoke them for a good 24 hours.

Our guru Keith Erlandson recommends exactly 36 hours at 24–26°C/75–80°F. Give them a little longer if it's cold outside and the temperature inside your smoker is rather lower.

Finish the breasts off by hot-smoking them for an hour or so at 110°C/230°F. If you vacuum seal them (see page 203), they will keep in the fridge for up to 8 weeks. If you merely store them in a sealed container, you should eat them within 7 days. They also freeze very well, like many smoked products, and will keep this way for up to 6 months.

Try either kind of breast thinly sliced in a salad with beetroot, new potatoes and flat-leaf parsley.

GOOSE BREAST WITH TRUFFLE OIL

Hot-smoked goose breast is indescribably good. It has a natural affinity with another delectable item, white truffle oil (see page 132). Nick told Johnny that he had cooked this delicacy as a gourmet starter for a posh dinner party. 'Why wasn't I invited?' demanded your ever-hungry author. 'Because I only use you for experimentation purposes,' Nick reminded him. Serves 4 as a starter

150g (5oz) Jerusalem artichokes, cleaned and sliced very thinly with a mandolin or food-processor
Vegetable oil
A couple of handfuls of wild rocket
1 medium-sized hot-smoked goose breast (see left), very thinly sliced

28 fresh sage leaves, deep fried until crispy (about 1 minute)
12 walnut halves, baked in the oven for 10 minutes at 180°C/350°F/gas mark 4
Parmesan shavings
White truffle oil (see page 132 – a tiny amount goes a long way)
1 teaspoon good-quality balsamic vinegar
cracked black pepper

- Making sure that the slices are so thin they are translucent, deep fry the artichokes in hot oil until golden and crispy. Do this in small batches. Drain on kitchen paper and reserve.
- Make a mound of rocket in the middle of a large, flat, white plate, then drape slices of goose breast around it, overlapping if necessary.
- Garnish with the rest of the items. You'll only need a few drops of truffle oil, as it's pretty strong stuff, plus about a teaspoonful of balsamic vinegar.

SMOKED VENISON

Smoked venison is a great delicacy. It is among the darkest of smoked meats and has a deep, sweet flavour. It is also very low in cholesterol. However, different cuts and species of deer require different treatment. This is a complicated business, so here we restrict ourselves to the hot-smoking of fillets of boneless haunch. These can be bought from specialist suppliers and some supermarkets. As with most hot-smoked products, venison needs brining, drying and a period of cold smoking before it is finished off in the hot smoker.

WEIGHING, BRINING AND DRYING
Get hold of a thick fillet of venison weighing 300–500g (10½-1lb 2oz). Immerse it in a 70 per cent brine solution for 45 minutes. This is made by adding 220g salt to every litre of water (or 4½oz to the pint). After brining, leave to drip-dry for at least 12 hours either on a rack or suspended by string.

SMOKING
Smoke the meat in in the cold smoker for at least a few days and preferably as much as a week at 15–24°C/60–75°F.

Before hot smoking, the venison should be rubbed with olive or sesame oil (the latter adds a nice nutty flavour). Smoke for 1 to 2 hours at 105°C/220°F.

MATURING AND EATING
Leave the venison to mature for at least a day (it will freeze for up to 6 months), then carve it very thinly across the grain. It is good with hard-boiled eggs and beetroot salad.

HOT-SMOKED MACKEREL

Mackerel are treated with a certain amount of scorn in Europe, largely because they are at their best for only a few hours, which is a lot longer than it takes them to make their way to the average fishmonger. Nevertheless, their natural oiliness makes them superb candidates for the smoker. As with those to be cooked in their fresh state, you want to select firm-fleshed, beady-eyed specimens that have been out of the sea for less than 24 hours. Better still, catch and smoke your own.

Gut the mackerel thoroughly, including the gills and the main artery that runs down the spine. Immerse the fish in 70 per cent brine (220g salt per litre of water/4½oz per pint) for an hour, then dry them for 4 hours in a cool, airy place. Hang them with loops of string either threaded through the eyes (being dead, they won't mind) or the now empty gill-plates.

Cold smoke the mackerel for 8 hours at around 25°C/75°F, then finish them off with 1½ hours in the hot smoker at 95°C/203°F. You can store your mackerel in the fridge for up to a week or for up to 6 months in the freezer. They are wonderful with creamy horseradish (see page 69). For a nutritious brunch or breakfast, try serving them with omelettes.

SMOKED MACKEREL SOUFFLÉ

Nick enjoys sitting at vantage points staring vacantly out to sea through binoculars. On the whole this is a sad part of his life, because nothing ever seems to happen out there. But once, and only once, it did. He was perched on a crumbling cliff in Kent, overlooking one of the busiest shipping lanes in the world, when mackerel started to burst out of the sea. There were millions of them, turning the water white with foam as they chased a gargantuan shoal of bait fish. Nick grabbed his rod, jumped into his inflatable kayak and only returned when he couldn't stuff any more fish into the boat. Later, he made a smoked mackerel soufflé to celebrate. **Serves 4**

400ml (14fl oz) milk
40g (1½oz) butter
40g (1½oz) plain flour
50g (2oz) cheddar cheese, grated
6 eggs, whites separated from yolks
Pinch of grated nutmeg
Salt
¼ teaspoon ground white pepper

1 teaspoon paprika
1 tablespoon chopped flat-leaf parsley
1 tablespoon horseradish sauce (see page 69)
200g (7oz) smoked mackerel (see left), filleted and flaked
large soufflé dish with a capacity of around 1½ litres (2⅔ pints)

- Preheat the oven to 180–190°C/350–375°F/ gas mark 4–5.
- Butter the soufflé dish.
- Heat the milk in a pan, while in another melt the butter, add the flour and stir into a paste. Gradually add the hot milk, whisking continuously to create a smooth sauce. Add the cheese and stir until it has all melted in.
- Remove the sauce from the heat, then add the egg yolks, nutmeg, salt and pepper, paprika, parsley, horseradish and mackerel. Mix thoroughly.
- Whip the egg whites up into soft peaks. Add them to the sauce a little at a time, folding them in gently to prevent them losing their airiness.
- Spoon the sauce into the soufflé dish and bake in the oven for 20 minutes, or until risen and golden brown. Serve immediately.

SMOKED TROUT

Johnny frankly won't touch trout unless it has been smoked. Nick likes the challenge of creating recipes that defy its customary blandness, but admits that it doesn't taste of much except mud in its natural state. But both agree that smoked trout is one of the world's great delicacies. Anthony Blunt, the royal art expert who was revealed in the 1970s to have been spying for the Soviets for years, agreed entirely. When he faced the media after his exposure and was asked what he intended to do now, he replied, 'I'm looking forward to some smoked trout.'

Both brown and rainbow trout are suitable for the hot-smoker.

WEIGHING, BRINING AND DRYING
First gut the fish and remove the main blood artery that runs the length of the spine, then weigh. Soak in the now-familiar 70 per cent brine (see page 77) for 1 hour, then thread string through the gill plates and hang the fish to dry for 2 hours, inserting short lengths of soaked cocktail sticks or matches (having first removed the ignition tips) inside the belly to hold the flaps open.

SMOKING
Cold-smoke for 8 to 10 hours and then hot smoke for 1½ hours at about 82°C/180°F, and certainly not more than 90°C/195°F. The fish should have lost about 15 per cent of their weight compared with how it stood immediately after gutting.

SMOKED TROUT KEDGEREE

This is actually somewhere between a risotto and a kedgeree. Nick was a bit hazy about how to make the latter, which usually features smoked haddock, so he decided to reinvent the wheel. The result is a delicate and classy comfort dish, excellent for a late weekend breakfast. Serves 4

2 shallots, sliced thinly
2 cloves of garlic, chopped
2 teaspoons mustard seeds
50g (2oz) unsalted butter
300g (10½oz) risotto rice
600ml (1 pint) chicken stock
4 cloves
A pinch of saffron strands
150g (5oz) fine asparagus, sliced
300g (10½oz) smoked trout fillet (see left), flaked
2 hard-boiled eggs, chopped
150ml (¼ pint) crème fraîche
Salt
A pinch of white pepper

- Fry the shallots, garlic and mustard seeds in the butter over medium heat for 4–5 minutes, until they start to colour slightly.
- Add the rice and stir well. Pour in the chicken stock, then add the cloves, saffron and asparagus. Simmer until all the stock has been soaked up and the rice is *al dente*.
- Add the trout, egg and crème fraîche and stir in well.
- Season with salt and white pepper and serve immediately.

HOT-SMOKED MUSSELS AND OYSTERS

Both these shellfish, when hot-smoked, are fantastic spread on hot buttered toast. Soft, creamy and seaweedy green in places, their fishy essence shines through the smokiness. And for once you don't have to bother with a preparatory period of cold-smoking.

PREPARATION
You need to start with live bivalves. Scrub them thoroughly, removing the beards from the mussels and steam them in a steamer for 10 minutes. All the shells should have opened. Discard any that haven't.

BRINING AND DRYING
Remove the flesh from the shells, inspecting for and removing any grit. Dip them in 70 per cent brine (see page 77) for 30 seconds or less, then place them on reasonably fine-meshed racks to drip-dry for half an hour in a cool, airy spot.

SMOKING
Oysters should be hot-smoked on racks at about 85°C/185°F for 20–30 minutes depending on their size. Turn them over half way through to ensure they are evenly smoked. Do the same with mussels, which will only ever need to be smoked for around 20 minutes.

STORAGE
Store your smoked oysters or mussels in the fridge in an airtight container for up to 1 week.

SHRIMP AND SMOKED OYSTER SKEWERS

We use tiger prawns for this refreshing summer snack, but any large, meaty crustaceans will do. The recipe works best with lightly smoked oysters. An alternative approach is to use raw ones. In this case, first thread the ingredients onto the skewers, then pop the whole lot into the hot smoker for a few minutes before finishing off on the barbecue or char grill. Naturally the end result will be smokier this way, or at least the prawns and courgettes will be.

You will need quite long skewers or kebab sticks for this, approximately 30cm (12in) not including the handles. Serves 2

1 courgette, sliced	2 cloves of garlic, chopped
6 large oyster mushrooms	Juice of ½ a lemon
6 large raw prawns	A pinch of dried
6 large lightly smoked	Mediterranean herbs
oysters (see left)	Salt
2 tablespoons olive oil	Pepper

- Mix all the ingredients carefully in a large bowl, taking care not to damage the oyster mushrooms.
- Thread the solid items onto the skewers, firmly pressing them against each other.
- Cook on the barbecue for 4–5 minutes on each side until nicely charred.
- Slide the vegetables off the skewers and serve with rice or pitta bread and chilli sauce.

SMOKED MUSSEL AND CHEDDAR PIZZA

We call this dish pizza, but there are certain ingredients you'd be unlikely to find alongside one another in Italy. Oh well, one day the Italians may begin to appreciate the wonders of modern British cuisine. Maybe they'll start with this pizza. **Serves 2–4**

THE DOUGH
450g (1lb) '00' grade plain flour
2 teaspoons salt
1 teaspoon dried yeast
2 tablespoons olive oil
300ml (½ pint) water

THE TOPPING
150ml (¼ pint) tomato passata (see page 195)
60g (2½oz) semi-dried tomatoes (see page 28)
200g (7oz) mature cheddar, grated
2 tablespoons finely diced pancetta-style bacon (see page 70)
40 smoked mussels (see page 81)
1 teaspoon paprika
A couple of leaves of tarragon, removed from the stalk
Freshly ground black pepper

- First, make the dough. We always do this by hand because it is easy and fun; however it can just as easily be made in a food-processor. Sift the flour into a large bowl, then mix in the salt and yeast.
- Stir in the olive oil and water with a large spoon. Turn the sticky mass out onto a floured surface and knead until it becomes smooth. Place in a bowl in a warm place with a damp cloth on top. Nick places his by the side of the Aga. An airing cupboard will do or just on the side in a warm kitchen.
- After 1 hour the dough will have risen. Knead it again for 1 minute, then divide it in two balls. You will just need one for this recipe. The other one can go in the freezer.
- To make the topping, blend the passata and semi-dried tomatoes and gently boil for 15 minutes until reduced and thickened.

- To assemble the pizza, first preheat the oven to 230°C/450°F/gas mark 8.
- Roll out the pizza dough gradually, using a little flour to prevent it sticking. See the Pissaladière recipe on page 31 for your options concerning shapes and trays.
- Spread the tomato mix thinly onto the dough, then sprinkle the cheddar on top.
- Add a thin layer of pancetta, the mussels and a dusting of paprika.
- Place the pizza in the hottest part of the oven and cook for 10–15 minutes until brown and bubbling.
- Garnish with tarragon and black pepper and serve.

CRISPY SMOKED OYSTERS WITH SESAME SEEDS

If you are wary of raw oysters, and some people do find them a bit full-on, the solution may be to smoke them. This preserves their intense, slightly metallic fishiness but transforms their texture from slippery and elusive to firm, soft and melting. They are excellent with hot buttered toast, but also respond well to oriental treatment. **Serves 2**

THE DIPPING SAUCE
2 teaspoons honey
2 teaspoons toasted sesame seeds (dry-fried over medium heat until lightly coloured)
Juice of 1 lime
1 tablespoon sesame oil
1 tablespoon tamari
2 teaspoons finely chopped ginger
2 spring onions, chopped
1 sprig of coriander, chopped

THE OYSTERS
Vegetable oil, for frying
8 lightly smoked oysters (see page 81)
2 tablespoons plain flour
1 egg, lightly beaten
50g (2oz) medium breadcrumbs
1 tablespoon sesame seeds

- To make the dipping sauce, combine all the ingredients and whisk together vigorously.
- Heat the oil to 180°C/356°F in a pan or deep-fat fryer.
- Coat the oysters in the flour, then dip in the beaten egg. Mix the sesame seeds up with the breadcrumbs and coat the oysters in this mixture.
- Fry the oysters in the oil until golden, then serve immediately with the dipping sauce.

SAUSAGES

The word 'sausage' is derived from the Latin *salsus*, which means 'salted'. But sausages have been around for much longer than this implies. The Sumerians were making them almost five thousand years ago in Mesopotamia. A couple of millennia later, Homer mentioned

'These goat sausages sizzling here in the fire

we packed them with fat and blood to have for supper.

Now, whoever wins this bout and proves the stronger,

Let that man step up and take his pick!'

Chances are that the original sausage was a version of haggis, in which a sheep's stomach was stuffed with its chopped offal along with oatmeal and seasonings, then cooked. Quite wrongly, this dish is assumed to be Scottish in origin. In fact, the word 'haggis' was in use in fifteenth-century England, and it may be derived from old French. It is nearer the truth to say that the Scots are the only people who have yet to give the habit up. This is inexplicable in our eyes, as the 'great chieftain o' the puddin' race' is extremely good.

The preserving concept behind the sausage is that the skin, usually made from cleaned animal intestine, acts as a barrier against airborne organisms. In many varieties it is also treated in some way (for example, with oil, ashes, herbs or pepper) as a further deterrent. Meanwhile, on the inside, seasonings and spices carry on the antibacterial work, sometimes aided by natural fermentation, as with salami. Finally, sausages are often smoked, or air-dried where conditions allow, further prolonging their shelf-lives. The great British banger, untreated and originating in a damp climate, scarcely qualifies as a preserved food as it is not long- lived.

Without the sausage, the British would have been denied endless sit-com jokes, not to mention the full English breakfast. The kings in this field, however, are undoubtedly the Germans, who have a sausage for every occasion. They can be divided into four categories: cooked, raw, frying and 'miscellaneous'.

An idea of the enormous variety out there can be obtained by drawing up a sausage A to Z. Here's ours: Andouillette, Black Pudding, Chorizo, Drisheen, Epping, Frankfurter, Garlic, Haggis, Indiana, Jagdwurst, Kielbasser, Lap Chong, Merguez, Neopolitan, Oxford, Pepperoni, Quorn, Reinsdyrpolse, Salami, Toulouse, Ukrainian, Vienna, Weisswurst, Xtrawurst, Yershig, Zampone.

THE PRINCIPLES OF SAUSAGE-MAKING

If you get it right, sausage-making is among the most satisfying of all food-preserving techniques. It's quite a buzz to create something at home that wouldn't look out of place in a smart deli, particularly when it tastes better than anything you'd be likely to find in one. And making sausages is an engaging business right from the preparatory stages to the eating. Then comes the bitter-sweet, resolve-testing period of waiting for them to mature. But if you endure to the end, your patience will be rewarded.

So, to paraphrase Shirley Conran, life's too short not to stuff a sausage.

A PASSION FOR HYGIENE

Johnny is frankly neurotic about sausages. Part of this is an occasionally justified paranoia about them being used to 'launder' dubious ingredients (one of the big advantages of making your own is that you can control this side of things). The rest comes from spending time with his friend Marcus. Marcus is strangely cursed in the sausage department. During the Ceausescu era, he found himself trying a particularly rancid specimen in front of a Rumanian who had just spent several days' wages buying it for him. When he guiltily admitted defeat, the Rumanian wolfed it down in a fury, then became seriously ill. It was in the company of the same Marcus that Johnny once ate an undercooked banger on the coast of Northumberland, England. 'Is it meant to be like this?' he asked the waitress, sweetly imagining that blood-pink sausage might be a local delicacy. 'No, pet,' she'd replied, 'it just isn't cooked properly!' Days of distress and self-recrimination followed as Johnny prepared for the onset of food poisoning.

A little squeamishness is no bad thing when it comes to making sausages. Hygiene is incredibly important. Mincing meat greatly increases its surface area, hence the scope for microbial contamination, and mixing it will spread anything undesirable through the whole batch. You should therefore always work with chilled meat, which is easier to handle and minimises the chances of infection. You must also make sure that everything it comes into contact with is scrupulously clean. If you are making raw sausages, make sure you cook them thoroughly before eating them. If you are making the dried or semi-dried type, bear in mind that they will be hanging around in the open for some time. It is vital that you add the right preservatives to keep them safe during this period, and sometimes a starter culture is called for to direct the microbial saga that unfolds. Then you need to keep a careful eye on conditions while your sausages mature. If they get too hot or damp, they may go bad.

Sausage-making, as you may be gathering, is quite an undertaking. Your pride when you cut open a perfectly matured salami will reflect the love and attention you have lavished on it. You will appreciate it all the more for knowing how easily things could have gone wrong. And if you are still nervous, reassure yourself with the thought that everything in this chapter has passed the test of Johnny's anxious and discriminating palate.

All we can do here is scratch the surface of what is really a specialist subject. With any luck, the experience of successfully making one or two of the delicacies below will inspire you to explore the world of sausages more fully. If you get serious about it, you may need to invest in some dedicated equipment. In the meantime, once you've got hold of casings, you can get by with rudimentary household kit, though it doesn't half help if you've got a mincer or food-processor with a sausage-stuffer attachment.

CASINGS

You have two main options as regards skins for your sausages. The traditional approach is to go for natural casings, which are made from the cleaned intestines of cows, pigs or sheep. They need quite a lot of preparation, but this only adds to the hands-on fun. They also have many advantages, being smoke-permeable, relatively elastic and also edible (although ox casings can be tough and are often peeled away before eating). The alternative is to use man-made casings. Reconstituted collagen ones are uniform, ultimately animal and come in a useful range of smaller sizes, but they are less flexible than natural casings. You also don't want to get them wet, a reversal of the way of things with the natural kind. Fibrous and plastic casings are popular with commercial manufacturers, but you can't eat them and we've decided they are beyond the scope of this book. We'd say go for natural every time.

Getting hold of sausage casings is easier than you might think. If you type the words into a search engine along with the name of a big local city, you are unlikely to draw a blank. Or get a friendly butcher to point you in the right direction.

Natural casings

The most useful natural casings for the home sausage-maker are sheep casings, hog casings and beef middles. Sheep casings are the thinnest and most delicate, with a diameter (i.e. thickness) of 16–26mm (⅔–1in). They are perfect for small, fresh 'cooking' sausages but too thin for dried ones. Hog casings are next up in width. They range from 32–45 mm (1¼–1¾in) in diameter and are good for fresh and dried sausages alike. To make a thicker salami or garlic sausage, you are better off with beef middles. These can be bought in diameters from 45–110mm (1¾–4¼in) and over.

Whichever kind of casing you use, it will need to be untangled from the bundle it arrived with, then cleaned and soaked. To do this, take more than you think you'll need, put them in a bowl and soak them in cold water for a few minutes. Drain and refill the bowl, then pour a few inches of water into a scrupulously cleaned sink and transfer the casings into this. One by one, flush them out by filling them with water from the tap. This will reveal any leaks. Then leave the casings in the sink, only taking them out as you stuff them.

MINCING

The meat in some sausages needs to be very finely minced, in others it can be coarsely chopped. Either way, a mincer of one kind or another is invaluable. You can buy a traditional cast-iron mincer with a range of cutting discs and sausage-filler attachments for less than £40. In the short run, a good butcher will do the job for you.

CURING

The question of how much to cure your sausages and with what is a fraught one. With fresh varieties this is less of an issue because the curative powers of salt, herbs and spices are sufficient for their short shelf-lives. With dried sausages, which need more protection due to their longer lifespan, curing is very important indeed. Health concerns, however, pull us in opposite directions when we ponder how to go about it.

On the one hand, we are increasingly uncomfortable about the role of chemical preservatives in our food. The ones most frequently employed in commercial sausage manufacture, like potassium nitrate (saltpetre), sodium nitrate and sodium nitrite, have had a bad press in recent years. On the other hand, sausage-makers have been using saltpetre, for example, for so long that it is now traditional. We have come to expect what it does to the taste and appearance of our sausages. And with the risks of contamination real, it pays to err on the side of caution when it comes to inhibiting the growth of bacteria.

Chemical caution

As a result of these contradictory pressures, commercial sausage-makers in many countries are both obliged to employ nitrates and/or nitrites and strictly limited in the amounts they can use. This is the kind of middle path we would advise you to take, using chemical preservatives where prudent but being very careful about quantities.

If this makes you anxious, some thoughts may put things into perspective. First, the lethal dose of potassium nitrate is more than an ounce (28g), which you would find difficult to ingest if you tried. Secondly, you do not want to risk poisoning yourself and others when you could have taken steps to prevent it. Finally, salt is ultimately as much of a chemical preservative as saltpetre, and there is no way round using a fair amount of that when making sausages. However, one rock is not the same as another, and we've restricted ourselves to two items which call for saltpetre or related curing agents on account of their long maturing periods. These are paprika salami and garlic sausage. You can make these without the curing agents, but let's just say it's safer not to.

A convenient way to feel more confident about preservative chemicals is to a purchase ready-made cure mix from a butchers' supply firm (see Useful Addresses on page 218). Prague powder number 2 is a popular mix for dried sausages, but there are others. All should come with detailed instructions about quantities to use.

FERMENTING AGENTS

Some of the best sausages owe their texture and flavour to internal fermentation. More information on the subject can be found in a later chapter, but the crucial bacteria to encourage for properly matured sausages are the kind that produce lactic acid. This discourages unwanted bacteria and lends the finished products a unique flavour.

In the old days, people relied on chance to bring desirable micro-organisms to maturing sausages, enhanced by preparing them according to procedures that had worked before. Once a successful specimen had developed, they found they could use it to 'infect' younger sausages by hanging it close to them, though they often had no idea of the mechanism.

Starter cultures

The more reliable modern technique is to use a starter culture to get the fermentation process going. You can buy these freeze-dried from butchers' suppliers (see Useful Addresses on page 218). They will keep in the freezer for 6 months. The easiest lactic-acid-producing culture to obtain is probably acidophilus powder. This can be bought in health food shops and then ground up with a pestle and mortar if sold in tablet form. Two of our recipes (paprika salami and garlic sausage again) contain acidophilus, although you can substitute one of the commercial cultures if you prefer.

You will need to mix your starter culture with a little warm water before incorporating it into the mix. To kick it into action, hang your dried sausages in a warm place for 24 hours (30°C/ 86°F) before moving them to a cooler environment to mature.

STUFFING

The exact mechanics of stuffing your sausages will depend on your equipment. You can stuff a sausage with a piping bag, or even with your hands at a pinch, but a mechanical device with a stuffing attachment is easier (you can buy such an attachment for your food-processor). At all events, you will probably end up tying a knot in one end of a length of casing while sliding the other onto a greased nozzle. Push some of the slack casing onto the nozzle so it bunches up concertina-style. Then start to feed in the meat. If you are using natural casings, keep them wet throughout. It is also a good idea to have a bowl of water nearby to rinse sticky fingers.

As the meat goes into the casings, manipulate it so that it is firmly packed the whole length of the sausage. Don't over-stuff or the skins may rupture. What you do when your sausage is the right length depends on the variety. If you are making fresh links, twist the casing and move on to the next one. If you are making dried sausages, cut the casing near the nozzle and tie it in a knot. Then tie another knot just below the first one with a length of string, leaving one end long for hanging purposes.

HANGING AND STORING

When you examine your just-stuffed sausages, you may need to prick the casings here and there with a pin or knife-tip to ensure there are no air pockets.

Fermented sausages need to hang in a warm place for a day or so to get the process going. A drying box (see the biltong section on page 14) is ideal for this. Thereafter they should be moved somewhere cool to mature. Other kinds cut straight to the cool phase.

Almost anywhere will do so long as it is airy and the temperature doesn't rise above 15°C/60°F. In a European winter, any porch or garden shed should be suitable. An easy way to provide a hanging rail is to stretch a broom handle between two chairs, placing some foil underneath to catch drippings. To minimise the risk of cross-contamination, make sure no two sausages touch as they hang .

Cooking sausages can be frozen for up to 3 months. Dry ones will keep in the fridge almost indefinitely, but should be eaten within 3 weeks once you've cut into them.

PAPRIKA SALAMI

The Hungarians are the world masters of salami, flavouring it with paprika, which they are also big on. This gives the salami both sweetness and bite. Johnny once brought a magnificent example back from Budapest, but his ex-girlfriend inadvertently ruined it by popping it in the freezer. Hence the 'ex'.

1kg (2¼lb) pork shoulder, roughly minced
300g (10½oz) pork back fat, cut into small dice
1½ teaspoons paprika
1 teaspoon ground black pepper
35g (1¼oz) salt
½ teaspoon cayenne pepper
1 teaspoon caraway seeds
¼ teaspoon saltpetre, or commercial cure mix, consulting label for quantity
¼ teaspoon acidophilus powder, or other starter culture, consulting label for quantity
casings – 50-80mm (2-3in) beef middles or widest available hog

Mix the minced pork shoulder with the back fat, then add the paprika, black pepper, salt, cayenne, caraway seeds and saltpetre or commercial cure mix. Add the starter culture, having first dissolved it in 1 tablespoon of water, and then mix everything together thoroughly.

Stuff the mince into the casings. If you are using hog casings, the salamis will mature faster as they will be narrower, but they will be a little thin. You should aim to make them approximately 30cm (12in) in length.

The next stage is to incubate the salami. This entails hanging them in a warm environment (around 30°C/86°F) for 24 hours to get the fermenting process going. Your drying box (see page 14) has all the right credentials, but you may need to leave the door or lid slightly ajar to maintain the right temperature. Place an open container of water inside to keep the environment humid.

The salamis should now be cold-smoked for 2 days at a temperature not exceeding 22°C/72°F. Beech wood imparts a good flavour. They will turn a deep smoky orange.

Now you can hang the salamis. Pick somewhere cool and airy, where the temperature is unlikely to rise above 12°C/54°F, and leave them for at least 2 months if you used hog casings or 4 if you used beef. Inspect them from time to time during this period – if the meat in one seems a bit loose, you can compact it and re-tie the ends. Don't be surprised if white mould appears on the surfaces of the sausages during the maturing phase. It's meant to.

Your salamis are edible when they feel firm and look dry, but you can leave them hanging for longer if you prefer them hardish. When they reach a desirable condition, you can slow down their hardening by rubbing off their surface mould with a cloth and rolling them in wood ash.

Keep your salamis cool in a larder or fridge and eat within 1 month of cutting the first slices.

THE TAPLOW CURED SAUSAGE

Johnny lives in a village called Taplow on the Berkshire/Buckinghamshire border. It's a nice enough place, and he thought it high time it had its own sausage. So this is what we devised. The Taplow illustrates how easy it is to dream up your own sausage recipes. It is versatile – you can smoke it or not according to your inclination – but only slightly preserved, so you should eat it within a week of refrigeration.

1.5kg (3lb 5oz) fatty pork belly, roughly minced
1 tablespoon salt
5 cloves of garlic, chopped
2 teaspoons caraway seeds
1 teaspoon cracked black pepper
30ml (1fl oz) sherry vinegar
½ teaspoon ground cloves
1½ teaspoons dried sage
1 tablespoon angostura bitters
34mm (1⅓in) diameter hog casings

Mix the ingredients together and leave in the fridge overnight.

Prepare the hog casings by soaking them, then running water through them. Fill the skins with the sausage mix. You want the individual links to be 10–12cm (4–5in) long. Separate them as you make them by twisting the casings.

Leave the sausages to mature in a cool place for 24 hours, then hang them or place them on racks in a cool, well-ventilated room for another day.

At this stage, the sausages can be smoked if desired. They will need 6–12 hours in the cold-smoker.

The Taplow comes into its own fried and served with mashed potatoes and cabbage or in a sausage sandwich.

FRENCH-STYLE GARLIC SAUSAGE

This is an excellent introduction to the joys of fermented sausages. Tangy, pink slices of the finished product are great in sandwiches and salads, and the maturing phase is quite dramatic. From the fifth day onwards, the sausages will bloom with a white mould as it feeds off the lactic acid produced by fermentation. The key to the process is the acidophilus, which is a member of the valuable lactobacillus family. These 'friendly' bacteria are responsible for the flavour changes we treasure in cheese, beer and sour dough bread. Here we see what they can do for a sausage.

1kg (2¼lb) pork shoulder
300g (10½oz) hard pork back fat
35g (1¼oz) salt
1 teaspoon roughly cracked black pepper
1 teaspoon dried herbes de Provence (see page 20)
½ teaspoon cayenne pepper
¼ teaspoon nutmeg
30ml (1fl oz) brandy
5 cloves of garlic, finely chopped
¼ teaspoon acidophilus powder, or other starter
   culture, according to packet instructions
¼ teaspoon saltpetre
Beef middle casings

Mince the pork shoulder with a 'medium' cut or get your butcher to do it for you. Hand-cut the back fat into short strips approximately 20mm long and 5 mm wide (¾ by ¼in).

Place all the ingredients in a bowl and knead thoroughly.

Stuff the casings, making sausages about 30cm/12in in length and tie up the ends.

Hang the garlic sausages in a warm place for 48 hours, for instance, in a drying box (see page 14) or over the Aga.

Transfer them to a cool place and hang for at least 4 months. Nick puts them under the stairs where they mature at a constant 15°C/60°F. The optimum maturing/drying period is 6 months. The sausages will keep in the fridge for 3 months more, but eat within 3 weeks once you have cut into it.

CHORIZO

The Spanish are justifiably proud of their chorizo, but the Latin Americans know a bit about it too. This recipe comes from Mexico, where Nick was once accidentally thrown downstairs by a salsa dancer. Just before he passed out, he gazed up at a bunch of chorizo hanging from the ceiling. During his recovery, he fixated on learning how to make it.

This is not a particularly long-lived chorizo, and unlike some varieties it must be cooked before being eaten.

1.5kg (3lb 5oz) fatty pork belly, roughly minced
2 tablespoons picante pimenton (hot paprika made from smoked chillies)
1 tablespoon ground achiote (red seeds of the annatto tree)
50ml (2fl oz) sherry vinegar
3 teaspoons salt
1 teaspoon cracked black pepper
6 cloves of garlic, finely chopped
40mm (1½in) diameter hog casings

A trip to a Spanish or Latin supermarket should furnish you with the pimenton and achiote.

Mix the pork belly with the rest of the ingredients and leave in a cool place for 24 hours.

Fill the meat into the casings, making small sausages about 10cm (4in) long.

Hang the chorizo to dry in a cool, airy place for 48 hours.

Store in the fridge for up to 4 days or 6 months in the freezer.

For a satisfying Mexican breakfast, fry or char-grill your chorizo and serve them with *huevos rancheros* (eggs with salsa and grated cheese on soft tortillas).

CHARGRILLED CHORIZO AND SQUID SALAD

The combination of pork product with cephalopod is not immediately obvious, but the marriage is a happy one. In Galicia in north-west Spain, they have known this for a long time. Serves 2

1 red pepper
1 head of garlic
1 tablespoon balsamic vinegar
3 tablespoons olive oil
Salt
Black pepper
2 medium potatoes, cut into 2cm (¾in) dice, then simmered for 10 minutes
½ aubergine, diced
A few sprigs of rosemary
150g (5oz) baby squid
Lemon juice
Paprika
150g (6oz) chorizo, sliced lengthways
Radicchio leaves

- To roast the red pepper, place in a hot oven (220–230°C/425–450°F/gas mark 7–8) for about 30 minutes until charred. Wrap in a plastic bag, then peel and de-seed when cold.
- To roast the garlic, take the entire head, slice off the top and place on aluminium foil. Dribble a little olive oil into the garlic, then close the foil over it. Roast alongside the red pepper for a similar length of time. Leave the garlic to cool, then squeeze out the sweet pulp by hand.
- Blend the red pepper in a food-processor with the vinegar, 1 tablespoon of the olive oil and a little salt and pepper.
- Sauté the potato cubes with the aubergine and rosemary in 2 tablespoons of olive oil over moderate heat until golden brown, then season with a little salt.
- Season the squid with the lemon juice, salt, pepper and paprika and char-grill it with the chorizo until both look temptingly crisp and crunchy.
- Mix together all the ingredients and season.

MARIA ROSA'S ITALIAN SAUSAGE

Nick has worked with Valentino for years. Every autumn, Valentino's mother, who lives in Southern Italy, makes sausages. They are smoked, flavoured with fennel seed and very good.

Maria Rosa uses roughly minced pork belly in her sausages with a fat content of approximately 35 per cent. This is what makes them so tasty. She uses hog or ox casings according to whim, giving a diameter of 40–80mm (1½–3in).

FOR EVERY 1KG (2¼ LB) OF PORK, YOU NEED
30g (1 generous oz) salt
1 tablespoon ground hot red paprika
½ tablespoon wild fennel seed

Knead the ingredients together. Add a couple of drops of water to loosen the meat and help release the flavours. Cover the mixture with a cloth and leave it to rest in a cool place for 12 hours (not the fridge). At the end of this period, knead the mixture again.

Prepare your casings (Maria Rosa squeezes lemon juice on hers), then fill the sausages. You want them each to be about 15–20cm (6–8in) in length.

Pierce each sausage in several places with a pin to allow the added water to run off. Then hang them horizontally in a cool, well-ventilated room for 24 hours.

Maria Rosa cold-smokes her sausage for 24–36 hours, then leaves them hanging in the smoker for another 20 days until 'ripe'. We suggest you do the same. The temperature should not exceed 15°C/60°F. After this, Maria Rosa cleans her sausages with a cloth, then vacuum seals them or stores them in olive oil.

These sausages are dark and strong. They should be sliced thinly and eaten raw with sour dough bread and olive oil.

ITALIAN SAUSAGE SALAD

This salad, filled with Italianate ingredients, gives Maria Rosa's cured sausage (see left) an excellent opportunity to show off its talents. It's ideal for *al fresco* eating. If there are a few of you bringing food to a picnic, this will inevitably be the star turn. Serves 4

250g (9oz) cherry tomatoes
Splash of olive oil
A little balsamic vinegar
1 medium red pepper
200g (7oz) cured Italian sausage, finely diced
150g (5oz) buffalo mozzarella, diced
100g (3½oz) shredded radicchio leaves
30 pitted black olives
100ml (3½fl oz) basil purée
Juice of ½ lemon
1 tablespoon balsamic vinegar
Salt
Freshly ground black pepper
8 slices of ciabatta, slowly toasted in the oven at 165°C/325°F/gas mark 3 until dry (15–20 minutes)

- Roast the cherry tomatoes in the oven for 30 minutes at 190°C/375°F/gas mark 5 with a splash of olive oil and balsamic vinegar.
- Then turn the heat up high (220°C/425°F/gas mark 7)and roast the red pepper for about 20 minutes. Peel, de-seed and slice it, using the 'plastic bag' method (see page 94) if you so choose.
- Combine all the ingredients except the ciabatta. Either serve the salad immediately or store chilled in an airtight container until needed.
- Serve with the toasted slices of ciabatta.

FRANKFURTERS

Although the citizens of Frankfurt-am-Mein threw a party in the 1980s to celebrate the 500th anniversary of their famous local sausage, the modern version appeared in Germany during the late nineteenth century. It was made of pork with a little salted bacon. Then the Americans adopted the sausage as their own and completely changed it, frequently making it from beef. Fortunately, we are in possession of the original recipe.

On the street, eating hot dogs is a hit-and-miss business, but you can rest easy if you've made your own franks.

800g (1¾lb) lean pork shoulder
150g (5oz) smoked bacon (see page 70)
1½ teaspoons salt
½ teaspoon ground white pepper
½ teaspoon ground mace
1½ teaspoons paprika
½ medium onion (about 50g/2oz)
75ml (3fl oz) water
1 teaspoon corn flour
34mm (1½ in) diameter hog casings

Mix all the ingredients apart from the onion, water and corn flour and blend it in a food-processor with the blade attachment until smooth. You may need to do this in batches. Leave the meat to rest for a few hours in the fridge.

Purée the onion in the food processor, then heat it up in a small saucepan with the water and corn flour. Stir constantly. As soon as the corn flour thickens, remove from heat. Allow to cool, then add to the meat and mix in thoroughly.

Prepare the hog casings, then fill the sausages. We like them long; the only limit is the size of your frying pan.

To cook the frankfurters, heat a large pan of water or beef stock and simmer for 20 minutes. Do this immediately.

You can remove the sausages at this stage and store them in the fridge for up to 5 days. Alternatively, make like the Germans and serve them in their simmering liquor with sauerkraut, sweet mustard and a hunk of bread. Any that remain can be fried later and served in rolls with grilled onions and ketchup.

If you vacuum-seal your frankfurters (see page 203), they will keep in the fridge for 3 weeks. You can also freeze them for up to 6 months.

HOT DOGS

Fry or barbecue some franks for 4–5 minutes or until starting to brown. Cut open the same number of long, soft rolls and shove a frankfurter into each. Add a dollop of sauerkraut (page 144), a streak of ketchup (page 102) and perhaps some sweetcorn relish (page 116). Eat with several paper napkins to hand.

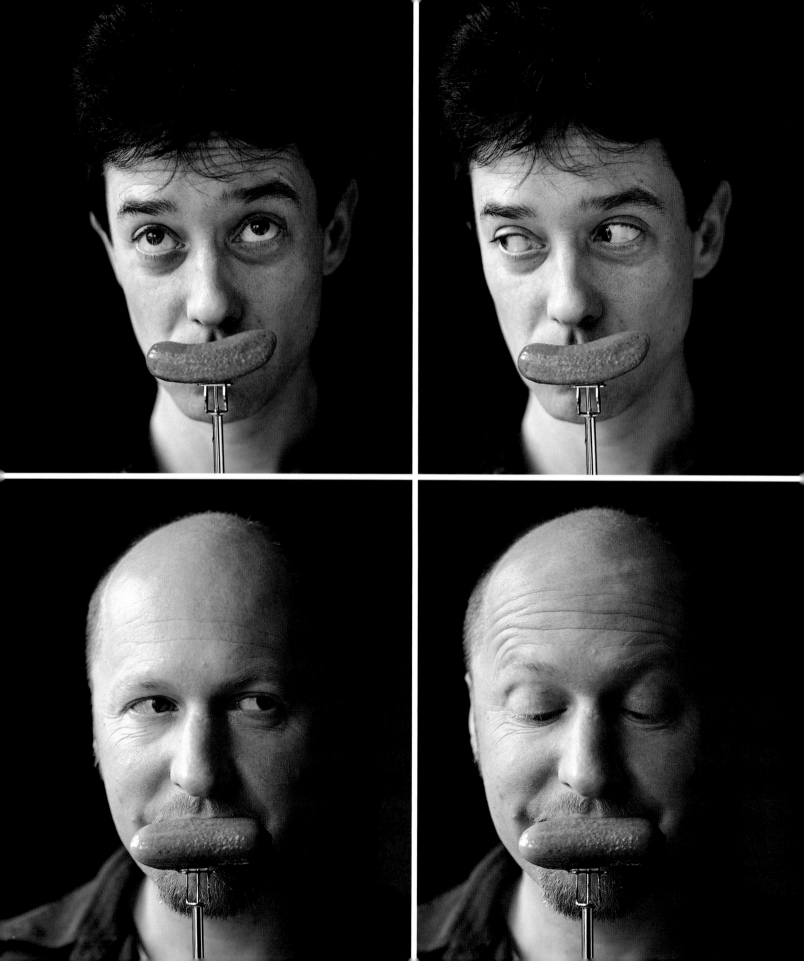

PICKLING

The word 'pickle' has reached us, via Low Medieval German, from *peik*, an Indo-European root meaning 'sharp-pointed'. And sharp is exactly how pickled foods taste, due to their high acid content. This is what produces that pleasurable shudder when you first bite into a cornichon or pickled onion.

Pickling works because microbes don't like acid. Although today the term is synonymous with long immersion in vinegar, the earliest pickles were made by fermentation (see Fermenting chapter). Bacteria occurring naturally on the surface of cucumbers, cabbages and other suitable vegetables would, given the right conditions, convert the sugars they contained into acid. This made them last a very long time.

Alcohol exposed to the air gradually turns into acetic acid, which is the active ingredient in vinegar. In medieval Europe, malt vinegar was produced in brewing areas and wine vinegar in the vineyards of the south. The Japanese, meanwhile, made a similar product from rice liquor. People soon discovered that vinegar had an excellent preservative effect on vegetables, and also that it kept them nicely crunchy. Verjuice, the unfermented but still acidic juice of grapes or apples, was also commonly used for pickling until the nineteenth century.

In England, pickled vegetables became seriously in vogue during the sixteenth century. Increasing affluence made beer and wine products more readily available, and salted foods began to be regarded, rather snobbishly, as linked with poverty. Onions, eggs and walnuts were the first big vinegar-pickled products, and the first two at least are still sold in many a traditional pub and fish-and-chip shop.

The most venerable pickled food, however, is the cucumber. That little slice of tangy green in your Whopper or Big Mac has a very long pedigree indeed. Cucumbers originated in India, and the word 'gherkin', which refers to small pickled ones, is derived from an Aryan word, derivatives of which are shared by people as diverse as the Czechs and the Greeks (not to mention the latest skyscraper in the City of London). The Mesopotamians were pickling cucumbers in brine 4,500 years ago. Aristotle praised their healing effects and Cleopatra considered them an important beauty aid. They were served at a feast thrown by King John of England in the early thirteenth century. More recently, the statesman Thomas Jefferson wrote that 'on a hot day in Virginia, I know nothing more comforting than a fine spiced pickle, brought up trout-like from the sparkling depths of the aromatic jar below the stairs of Aunt Sally's cellar.'

Pickles are good examples of foods that people originally grew to love through necessity rather than first impressions. Give a small child a pickled onion and you will see what we mean. The taste for this kind of food has to be acquired, but, once there, it is there for ever. Interestingly, the Chinese are just as convinced by the old wives' tale that pregnant women crave pickles as we are in the West.

TOMATO KETCHUP

Ketchup is now regarded as quintessentially American, but it was originally a Chinese and South-East Asian dipping sauce made from the pickling brine in which fish or vegetables had been preserved. In the seventeenth century, English sailors picked up on the delights of *ke-tsiap* and brought it back home with them. Europeans and Americans started to devise ketchup recipes based on liquor from their own favourite pickled foods, including mushrooms and walnuts. They also experimented with various ways of adding body to the thin oriental version of the sauce before hitting on the tomato in the mid-nineteenth century.

Tomato ketchup causes a great deal of hand-wringing among the middle classes, who view it as a kind of dangerous drug. They know that they love it, and their children still more so, but the depth of this passion makes them anxious. Were they to allow it free rein, they fear that all their food would end up smothered in ketchup. They are haunted by the thought of kids who can only be bribed to eat with tomato sauce. It is regarded as *declassé* and suspiciously multi-national.

Welcome to guilt-free ketchup. This version is less sweet than Heinz's and slightly thinner, depending how much you reduce it. It is more rounded and the taste of the spices comes through more clearly. Crucially, you will have made it yourself.

The tomatoes you use do make a difference. Make sure that they are ripe and sweet. Your average supermarket tomatoes are too watery. Cherry tomatoes are expensive, but they do make for fantastic ketchup. Otherwise, try to get hold of a job-lot of non-forced tomatoes from your local market at the end of the season (early autumn). Better still, grow and process your own.

TO MAKE ABOUT 1.6KG (3½LB)

3kg (6¾lb) tomatoes, chopped
150ml (¼pint) cider vinegar
10 cloves
4 cardamom pods
½ teaspoon white pepper
½ teaspoon ground mace
½ teaspoon ground allspice
½ teaspoon ground cinnamon
2 teaspoons paprika, medium heat
75g (3oz) white sugar
2 teaspoons salt
4 cloves of garlic, chopped
25g (1oz) sun-dried tomatoes (see page 28) chopped

Put all the ingredients into a large saucepan and bring to the boil. Reduce the heat and boil gently until the ketchup has reduced in volume by a third. This will take about 2 hours.

Pass the mixture through a food mill, then test it for thickness. If it isn't of the desired consistency, reduce it still further or thicken with a little corn flour.

Store the ketchup in sterilised bottles (page 164). If you are using bottles sealed with corks, see also pages 189–190 and dip the top of the bottle several times into melted candle or sealing wax, allowing each layer to set before re-dipping.

We would recommend a maximum pre-opening storage time of 8–10 months. Once you've opened a bottle, keep it in the fridge.

We doubt you really need tips on how to use tomato ketchup, but you could always try it with bacon and eggs.

MUSHROOM KETCHUP

Mushroom ketchup is considerably older than the tomato kind and brings an element of Victorian luxury to dishes that incorporate it. This recipe produces a sauce which is relatively thin compared with tomato ketchup. The port makes it rich and slightly sweet. It is, of course, intensely mushroomy. You can use it to great effect in a range of soups and sauces, in a modern stir-fry or an old-world steak-and-kidney pudding. We like it best, though, with char-grilled sirloin steak.

TO MAKE ABOUT 800ML (28FL OZ)

- 1.2kg (2lb 10oz) sliced mushrooms*
- 2 tablespoons sea salt
- 75g (3oz) dried mushrooms (see page 21), e.g. porcini, cleaned well to remove grit
- 4 shallots roasted with a head of garlic (see below)
- Olive oil
- 100ml (3½fl oz) balsamic vinegar
- 150ml (¼ pint) white wine vinegar
- 200ml (7fl oz) port
- 25g (1oz) sugar
- 3 fresh bay leaves
- 8 cloves
- ½ teaspoon ground white pepper
- ½ teaspoon chopped ginger
- ½ teaspoon ground allspice
- Splash of soy sauce

*You could use wild mushrooms such as porcini, horn of plenty or saffron milk caps, but shop-bought chestnut or portobello mushrooms can also form the basis of an excellent ketchup.

Mix the fresh mushrooms thoroughly with the sea salt in a large bowl. Cover with a cloth and leave for 24 hours in a cool place (but not as cool as the fridge).

Soak the dried mushrooms for half an hour in enough boiling water to cover them. Meanwhile, roast the unpeeled shallots with a small head of garlic and a dash of olive oil for half an hour at 200°C/400°F/gas mark 6, then peel and reserve.

Remove the rehydrated mushrooms with a slotted spoon and place them in a large saucepan. Carefully add the soaking liquor, taking care to leave behind any grit.

Add the remaining ingredients to the pan, bring to the boil and simmer for half an hour. Leave the contents to cool in the saucepan, then bring them back to the boil and simmer for another 30 minutes.

Pass the ketchup through a food mill or blender, pour it into a clean pan and bring to the boil yet again. Continue until the sauce has reduced to a volume of 800ml (28fl oz). This will be easy to gauge if you've previously made a mark on the outside of the pan corresponding with the surface of the same amount of water poured into it.

Pour the ketchup into sterilised bottles or jars via a sterilised plastic funnel – in other words, boil the funnel and bottles/jars for 10 minutes before use (see page 164). If you are using bottles with screw-on lids, fit the lids with cut-out card discs to prevent the ketchup coming into contact with bare metal. If you are using bottles with corks, refer also to pages 189–190. If you choose the cork method, dip the top of the bottle several times into melted candle or sealing wax, allowing each layer to set before re-dipping.

Leave the ketchup to mature in a cool place for at least 48 hours before use. It will have a shelf-life of at least 6 months and taste better as time goes by.

PRESERVED LEMONS

Silky preserved lemons are an essential ingredient in Moroccan and North African cuisine. They impart a fragrant, sweet yet sour taste to tagines and salads and go particularly well with lamb and chicken. Once pickled, you can eat the whole thing, rind and all.

TO FILL A 500ML (18FL OZ) PRESERVING JAR
 10 medium unwaxed lemons
 salt
 4 bay leaves
 15 peppercorns
 15 coriander seeds
 6 cloves

Cut half the lemons into wedges, 8 per lemon, then remove the seeds.

Squeeze the remaining lemons and reserve the juice.

Firmly press a layer of lemon wedges into the bottom of a sterilised jar (see page 164). Cover with 2 teaspoons of salt, a bay leaf, a few peppercorns and coriander seeds and 1 or 2 cloves. Press down another layer of lemon wedges and repeat the salting and spicing process. Continue until there is only 3cm (1in) or so of space at the top of the jar.

Now pour in the lemon juice, seal the jar (see page 164) and store under the stairs for 1 month before using (or somewhere else if you live in a bungalow). The lemons will keep for 1 year.

You may want to check the lemons every few days. Some batches let off a bit of gas and this will relieve the pressure. But this problem is unlikely to occur if you've used enough salt.

SLOW-COOKED LAMB WITH PRESERVED LEMONS

This recipe demonstrates what a little bit of age does to a salted lemon. The result is an eloquent advertisement for the transformative powers of pickling: the character of the fruit is changed beyond recognition. Because the stew is so fragrant and tangy, a little goes a long way. **Serves 4–6**

650g (1lb 7oz) lamb leg meat, cut into large dice

70ml (2½fl oz) olive oil

2 teaspoons ground cumin

1 teaspoon ground coriander

1½ teaspoons paprika

1½ teaspoons chopped garlic

2 medium onions, sliced

6 medium-strength whole dried red chillies (see page 26)

4 cloves

½ teaspoon ground cinnamon

550ml (19fl oz) tomato passata (see page 195)

50g (2oz) semi-dried tomatoes or a little less sun-dried tomato, roughly chopped (see page 28)

300ml (½ pint) water

2 whole preserved lemons or 12 wedges of preserved lemon (see page 104), assuming 8 wedges per fruit

A large sprig of mint, chopped

A large sprig of parsley, chopped

Salt

- For this recipe you need a robust pan with a capacity of 3 litres (5½ pints) and a snug-fitting lid. It must be suitable for heating in the oven.
- Fry the chunks of lamb over fierce heat in 1 tablespoon of the olive oil until browned. This will take about 5 minutes. Reserve.
- Pour the remaining oil into the pan and gently simmer the cumin, coriander, paprika and garlic for a few minutes. Then turn up the heat and throw in the onions and fry for 10–15 minutes until soft, stirring frequently.
- Add the lamb, red chillies, cloves, cinnamon, tomato passata, sun-dried tomatoes, water and preserved lemons. Bring to the boil, then turn down to a simmer.
- Add half the mint and parsley and a little salt, then place the lid on the saucepan and put in a low oven (150°C/300°F/gas mark 2) and leave it there for 3–4 hours, cooking gently.
- After you have removed the stew from the oven, garnish it with the remaining mint and parsley. Remove the chillies and the preserved lemon wedges; they have done their work and may now be discarded.
- Serve with rice or couscous.

PICKLED WALNUTS

Pickled walnuts were once deservedly popular. Like many pickled foods, they go very well with cheese, as their nutty sharpness counteracts its cloying tendency. They are also excellent with cold meats and feature in a number of old-fashioned stews and casseroles.

The following is adapted from a recipe from the Women's Institute, who have much collective wisdom in these matters. It calls for less brining than older techniques. As Johnny will confirm, having grown up with a large walnut tree in the garden, the juice from the fresh nuts is apt to stain your fingers nicotine-brown, so wear rubber gloves when you're handling them. In the UK, late June/early July is the time to pick walnuts in the desired 'green' state, before the hard nuts have formed inside the shiny cases.

4.5kg (10lb) green walnuts (n.b. don't try to remove the cases)

450g (1lb) salt, dissolved in 4.54 litres (1 gallon) water

THE SWEET PICKLING BRINE

1.8 litres (3 pints) malt vinegar

450g (1lb) brown sugar

1½ teaspoons salt

1 teaspoon black peppercorns

1 teaspoon ground allspice

½ teaspoon ground cloves

½ teaspoon ground cinnamon

1 tablespoon grated fresh ginger

Prick each walnut a few times with a fork. Place them in sterilised jars (see page 164) and cover them with the first, simple brine, i.e. not the pickling one. Leave overnight, then pour out and replace the brine. Repeat this pattern for 3 or 4 days.

Rinse the walnuts to remove excess salt, then place them on a tray and leave them to dry in an airy place for a day or two. They will turn black; fear not, they're supposed to.

Now put all the ingredients for the pickling brine in a pan together and bring them to the boil. Add the walnuts and boil on for 5–10 minutes.

Spoon the walnuts into jars, then cover with the spiced vinegar and seal (see page 164). Leave them to mature for 6 months before eating them. They will keep for a few years.

SOUSED HERRINGS

While the Dutch and Scandinavians make them into rollmops and eat them raw, the British, at least traditionally, prefer their herrings soused. That is, when they aren't making them into kippers (see page 66). Soused herrings are extremely refreshing on a hot summer's day and are dead easy to make. They will keep in the fridge for a week or so. Mackerel can also be soused in the same way.

4–8 herrings, depending on size
Salt and pepper
1 small onion, sliced into rings
6 peppercorns
4 cloves
1–2 bay leaves
4 sprigs of thyme
A few stalks of parsley
150ml (¼ pint) water
150ml (¼ pint) malt vinegar

Scrape the scales off the herrings, then clean and gut them, removing the heads, fins and bones but leaving the tails on. Pat dry and sprinkle with salt and pepper.

Roll the fish up from the head end (starting at the other end doesn't seem to work) and secure with cocktail sticks.

Place them in a shallow ovenproof dish and add the onion, peppercorns, cloves and herbs. Pour in the water and vinegar until the fish are almost covered.

Cover with foil and bake in the oven at 180°C/350°F/gas mark 4 for 45 minutes to 1 hour, or until tender.

Leave the fish to cool in the cooking liquid before you eat them. They are best served cold. Try them in a green salad with chopped hard-boiled eggs.

PICKLED HERRINGS WITH SOUR CREAM AND LUMPFISH ROE

Lumpfish are desperately ugly. They are covered with distended bumps, and the male has a huge sucker on his underbelly to attach himself to nearby rocks when guarding the fertilised eggs. They aren't great eating fish, but their pink eggs are salted and dyed black to make a good and widely available caviar substitute. They combine nicely with pickled herring, perhaps because of the common North Sea origin. Serves 2

4 pickled herring fillets (see left)
1 teaspoon orange zest
150ml (¼ pint) sour cream
2 teaspoons lumpfish roe
1 tablespoon roughly chopped dill
Paprika
Blinis (you can buy these ready-made in supermarkets; heat according to packet instructions)

- Arrange the ingredients on two plates and dust with paprika. The sharp taste of the herring is well suited to the blandness of the blinis. These are stereotypically served with smoked salmon, but this salad makes a refreshing alternative.

PICKLED ONIONS

Though not ideal if you are about to go on a date, pickled onions will awaken the most jaded of tastebuds. They will also make you ravenously thirsty, hence the traditional jar in old-time pubs. You can eat them on their own when you need pepping up or serve them with cheese and cold meat. Pickled pearl onions are mandatory with a cheese fondue.

The key to success is to use tiny onions and to make sure the vinegar is sufficiently sweetened. Some of our early batches were unpleasantly sour. After much experimentation, we've found the following recipe produces the sweetest, crispest onions.

1.3kg (3lb) small pickling onions
50g (2oz) sea salt
Several cloves
Several blades of mace
1–2 fresh chillies, halved
400g (14oz) sugar
1.2 litres (2 pints) white wine vinegar

Trim the tops and bottoms off the onions, but don't overdo it or they will disintegrate in their pickle. Leaving the skins still on, pour boiling water over them and leave to blanch for 20 seconds. Then cover with cold water and peel them under water. This will prevent the surfaces oxidising and toughening up.

Layer the onions in a clean bowl, sprinkling each stratum with salt. Cover with a clean cloth and leave overnight. The salt will draw out much of their moisture, ensuring a desirable crunch.

Next day, rinse them well and dry them as thoroughly as possible. Place them in sterilised jars (see page 164), with two cloves, a blade of mace and half a chilli in each.

Boil up the sugar and vinegar for 1 minute, then pour the hot liquid over the onions. Seal the jars (see pages 164) and wait 2–3 weeks before eating. They will keep for at least 6 months.

PICKLED OCTOPUS

'My brain fidgeted like an octopus in a jar of vinegar' (**August Strindberg**)

Octopuses ('octopi' is plain wrong, 'octopodes' is correct but pedantic) have a range of defences to prevent you from eating their sweet flesh. They can make themselves almost any shape they desire, they can squirt ink at you and, if all else fails, they have very powerful beaks. If you do manage to catch them, you have to take extreme measures to render them suitably tender. Greek Islanders are often to be found pounding them on the rocks or hanging them on clotheslines in the sun. This recipe is for the Greek version of the dish, known to the locals as *oktapodi toursi*. The acid in the pickling vinegar helps the tenderising process. The purplish spirals of protein-rich tentacle will melt in your mouth.

2x 110g (4oz) baby octopuses
150ml (¼ pint) olive oil
150ml (¼ pint) red wine vinegar
4 cloves of garlic
Salt and black pepper to taste
A few stalks of thyme
Lemon wedges

To clean the molluscs, pull off their tentacles (reserving them) and remove and discard their intestines and ink sacs. Then cut out the eyes and beaks and throw them away. Skin the octopuses, then wash and scrub them thoroughly to remove any sand. Being babies, they shouldn't be too tough, but if they are, give them a good pounding with a mallet.

Place the head and tentacles in a pan with 6–8 tablespoons water. Cover and simmer gently for 1–1¼ hours. Test the flesh with a skewer. It should penetrate easily but not too easily. We like our octopus *al dente*.

Drain off any remaining liquid and leave the octopuses to cool.

Cut the flesh into strips 1cm (½in) wide and chop the tentacles into bite-sized pieces. You can remove the suckers at this stage, but we like the sensation they make on the tongue.

Pack the meat loosely in a sterilised screw-top jar (see page 164). Mix the oil and vinegar together and pour into the jar, making sure the octopus is thoroughly immersed. Stir in the garlic cloves and seasonings and push the sprigs of thyme into the jar. Remove any air bubbles clinging to the glass with a flat plastic spatula.

Seal the jar (see page 164) and leave in a cool place for at least 4–5 days before eating. To serve, drain the octopus and accompany it with cubes of bread and lemon wedges.

Unopened jars will keep for 1 year. Once opened, refrigerate and eat within 1 month.

PICKLED OCTOPUS SALAD WITH ROASTED PEPPERS

Nick ate the best octopus of his life in the Seychelles. Johnny prefers the Sicilian and Greek ways with the mollusc. Either way, we are both suckers for octopodes. They are a delight to watch as they propel themselves around under the sea, with waves of colour scudding across their bodies in response to the surroundings. They are highly intelligent, and can easily work out how to remove a stopper from a jar to get at a lobster. But, sadly for them, they are also very good on the end of a fork. Serves 2

1 red and 1 yellow pepper, roasted and sliced
1 aubergine, roasted and sliced into chunks
1 tablespoon olive oil
Salt
Pepper
200g (7oz) pickled octopus, (see left) sliced into bite-sized pieces
20 basil leaves

THE DRESSING
100ml (3½fl oz) red wine
1 teaspoon sugar
1 tablespoon balsamic vinegar
25ml (½fl oz) olive oil

- To roast the peppers, throw them into a very hot oven (230°C/450°F/gas mark 8) for 20 minutes until charred on the outside. Transfer into a plastic bag and leave until cool. They should now peel easily.

- To roast the aubergine, meanwhile, halve it lengthways and place both halves on a baking tray, face-up. Pour the olive oil over the exposed surface of each. Season with salt and pepper, and roast alongside the peppers. The aubergine halves are done when they are soft and slightly wrinkled.

- To make the dressing, combine the red wine, sugar and balsamic vinegar in a saucepan and reduce heavily, boiling over moderate heat until syrupy. Continue until there is only 1 tablespoon of liquid left. Then whisk in the olive oil.

- Arrange the peppers, aubergines, pickled octopus and basil on two plates. Dribble dressing onto each salad and serve.

PICKLED QUAILS' EGGS

Pickled hens' eggs are a staple in fish-and-chip shops, but they aren't as good as these feisty, cajun inspired, lilliputian ones.

Quails produce mighty big eggs, given their size. They are wonderfully dinky but they can be murder to peel. Our advice is to start with the thick end where the internal air space resides. Soaking them for a period in vinegar as per the recipe below will make the task much easier.

48 quails' eggs
700ml (1¼ pints) white wine vinegar, plus
   some more for pre-soaking the eggs
4 teaspoons salt
3 teaspoons cayenne pepper
6 cloves of garlic, peeled
12 peppercorns
10 whole allspice
2 teaspoons yellow mustard seeds
4 cloves
2 bay leaves
2 chillies

Place the eggs in a saucepan and just cover with water. Bring to the boil and simmer for 3 minutes, then drain and transfer immediately into a bowl of cold water to halt the cooking process. Then move them to a bowl of vinegar and leave for at least an hour. The speckles will come off due to the acid but this is no cause for alarm. After soaking, squeeze the eggs gently and peel them, making sure you remove the membranes under the shells.

Place the peeled eggs in sterilised pickling jars (see page 164). Then bring the remaining ingredients to the boil in a large saucepan. Turn off the heat and allow the liquid to steep for a couple of hours. Then pour it into the jars, making sure all the eggs are covered. Wipe the rims and screw on the tops.

Your eggs will be ready to eat in about 2 weeks. They will keep for up to 1 year in the fridge or a cool larder.

PICKLED CUCUMBERS

We rhapsodised the pickled cucumber in the introduction to this chapter. It comes in endless forms, ranging from the hefty dill pickle to the tiny, eye-watering cornichon. Pickled cucumbers are so prevalent that they are often simply called 'pickles'.

Pickling varieties of cucumber are slightly different from the 'slicing' kind, typically being thicker skinned with somewhat bumpy surfaces. The formula below is Jewish in origin. You can see a version made with dill on page 56, in the photo accompanying our recipe for salt beef. The two items are age-old companions.

6 large pickling cucumbers
1.8 litres (3 pints) water
6 cloves
3 cardamom pods
1 teaspoon black peppercorns
1 teaspoon caraway seeds
3 bay leaves
3 sprigs of fresh tarragon
125g (4½oz) salt
300ml (½pint) cider vinegar

Slice the cucumbers lengthways, then de-seed them by scooping the centre out with a teaspoon. Cut the cucumbers into slices approximately 1.5cm (⅔in) thick.

Heat the water to boiling point with the cloves, cardamom, peppercorns, caraway, bay leaves, tarragon and salt. Leave to cool, then add the vinegar. Stuff the cucumbers into sterilised jars (see page 164), inserting a sprig of tarragon in each half way up.

Pour in the pickling liquid to cover the cucumbers and seal the jars (see page 164). Make sure the lids do not have exposed metal on the inside. If they do, cut out some card to cover it.

Leave the cucumbers to ferment for a month in a cool, dark place such as a cellar. They should now be firm and sharp-tasting.

An unopened jar of pickles will keep for 1 year or longer. Once opened, refrigerate and consume within 2 months.

PICKLED ROASTED PEPPERS

Pickled peppers not only provide useful material for tongue-twisters, they also look and taste stunning. The visual effect is enhanced if you pack them in clear jars in alternating layers of red, yellow and green. You can pickle any varieties you fancy, although the smaller ones are difficult to peel, especially chillies. When ready, pickled peppers have a lovely meaty texture. If you know anyone called Peter Piper, you have a duty to invite him round to share them.

TO FILL TWO 500ML (18FL OZ) JARS
  1.5kg (3lb 5oz) mixed peppers
  250ml (9fl oz) distilled malt vinegar (5 per cent acidity)
  180g (6½ oz) white sugar

Roast the peppers whole in a very hot oven (230°C/450°F/gas mark 8) for 30 minutes or until nicely charred.

Pop them into a plastic bag to cool. As they sweat inside it, the condensation produced will make them easier to peel.

When they are cool enough to handle, peel and de-seed them over a sieve perched on a bowl. This will catch the juices. Pack the peppers in sterilised preserving jars (see page 164), in monochrome layers if they are of different colours.

Heat the vinegar and sugar in a pan and add the pepper juices. If the peppers didn't yield much juice or if you failed to catch it, add up to 200ml (7fl oz) of water.

Pour the hot pickling liquid over the peppers and seal the jars (see page 164). Store in the larder for up to 1 year.

Pickled peppers are great in warm salads, for instance, with tuna and rocket.

ROASTED CORN RELISH

Like ketchup, corn relish is associated with fast-food and so suffers from an image problem among foodies. This version, however, is so ideologically sound that it doesn't even have sugar in it. It's also tangy, and the corn kernels pop between the teeth in a highly satisfying manner.

You can eat our relish with many things (mature cheddar and tomato sandwiches come to mind), but it is most at home on the back of a burger.

TO MAKE 600ML (1 PINT)
  5 corn-on-the-cob
  2½ teaspoons yellow mustard seeds
  1 tablespoon vegetable oil
  50ml (2fl oz) distilled malt vinegar
  300g (10½oz) cherry tomatoes, blended in a food-processor
  1 teaspoon salt
  Ground black pepper
  ½g/pinch of saffron strands
  Small bunch of spring onions, finely sliced

First peel the husks off the corn-on-the-cob, then bake them in a hot oven (230°C/450°F/gas mark 8) for 30 minutes or until brown. When they come out they will look slightly crinkly. The blast treatment will have caramelised some of their sugars.

Shave the corn kernels from the cobs with a sharp knife, avoiding cutting yourself, and reserve.

Fry the mustard seeds in the oil in a large saucepan until they begin to pop. Add the rest of the ingredients and turn up the heat. Boil, stirring frequently, until the liquid starts to thicken.

Spoon into sterilised pots and seal (see page 164). The relish will keep in the fridge for 3 months.

BEEF BURGERS WITH ROASTED CORN RELISH

Back in the early 80s, when the UK first began to awaken from its culinary slumbers, the most sophisticated thing you could eat was an 'authentic' American burger. The peak experience was the arrival of the relish tray. The mandatory central trinity was something green (sweet cucumber pickle), something red (ketchup) and something vividly yellow (sweetcorn relish). The last one was always the most popular.

We make our burgers from pure, top-quality beef. No adulterants and no messing. **Serves 1**

175g (6oz) minced beef, with a fat content of at least 14 per cent

Salt and pepper

1 large burger bun or mini ciabatta, sliced lengthways

1 tablespoon mayonnaise and 1 tablespoon mild mustard, mixed together

roasted corn relish (see page 116)

4 slices of red onion

2 slices of beef tomato

A small handful of shredded iceberg lettuce

- Shape the beef into a nice, firm burger. Season with a little salt and pepper and char-grill to your taste. Alternatively, you could fry it.
- Heat the bun in a toaster or oven.
- Spread the mustard-mayonnaise mixture on the bottom bun, then lay the burger on top.
- Spoon the relish on top of the burger, then follow with the onion, tomato and iceberg lettuce. We expect you're familiar with the general procedure.
- Serve with fries and a side salad.

PICCALILLI

A recipe published in England in 1694 teaches the reader 'To pickle lila, an Indian pickle'. This is worrying from Johnny's perspective, as Lila is the name of his cat. Other readers may also have had alarming experiences with piccalilli, as shop-bought varieties are not always that great. This version, however, is crunchy with a bit of a kick. It helps explain how this bright yellow condiment got to be so popular, particularly with cheese.

We found ourselves truer to the Indian origins of this dish than we had intended when we couldn't get any pickling cucumbers on the day we planned to make it. Instead, we purchased some unfamiliar vegetables called parval and tindori, reassured by the local Indian grocer. He turned out to have given good advice.

TO MAKE THREE 300ML (½ PINT) POTS

- 1.2kg (2lb 10oz) mixed vegetables, diced into 1cm (½in) cubes*
- 2 tablespoons salt
- 1 tablespoon turmeric powder
- 50g (2oz) mustard powder
- ½ teaspoon ground white pepper
- 2 teaspoons ground ginger
- 50g (2oz) plain flour
- ¼ teaspoon ground nutmeg
- 5 tablespoons cider vinegar plus another 150ml (¼ pint) cider vinegar
- 250ml (9fl oz) malt vinegar
- a dash of water

In a very large saucepan, mix the turmeric, mustard, white pepper, ginger, flour, nutmeg and the 5 tablespoons of cider vinegar into a smooth paste. Slowly whisk in the rest of the vinegar, then add the vegetables, the malt vinegar and the water.

Gently heat the mixture until the sauce thickens. This should take 10–15 minutes. You must stir constantly or the flour will stick to the bottom of the pan. Remember that you don't want to actually cook the vegetables or they will lose their crunch, so take care not to over-boil.

Store your piccalilli in sterilised jars (see page 164). It will keep for 6 months in a cool larder and improve with age. Once opened, refrigerate and eat within 6 weeks.

Serve with roast chicken or frankfurters.

*We use carrot, cauliflower, shallot, turnip, baby corn, parsnip and pickling cucumber. Cauliflower is essential in anything worthy of the name 'piccalilli'.

Place the diced vegetables in a large bowl and mix them with the salt. Cover and leave overnight. The salt will draw out their liquid, giving them bite and intensifying the flavours.

HERBS, PASTES AND INFUSED OILS AND VINEGARS

Humans have treasured herbs and spices since prehistory. Initially this was largely on account of their medicinal and preservative properties, but as time has worn on, taste alone has become the primary motive for using them. This is particularly true in parts of the world where more sophisticated drugs and preserving methods have taken over.

But the tastiness and health-enhancing properties of herbs and spices are not unrelated. The human body knows a good thing when it smells or nibbles it. And as we are starting to see, a liking for foods that were once vital to survival tends to be passed down within a culture even when their original *raison d'être* has long-since lapsed.

Herbs can of course be dried, but another way of using them is to combine them with oil or vinegar. Depending on the ratio of the flavouring ingredient to its medium, this can produce a delicate infusion or a pungent paste. Often the original colours of the additives are retained, which is why pesto and mint sauce are so vividly green, and why glass-bottled infusions are so aesthetically pleasing.

Pastes, often very strong ones, are particularly important in hot areas like India, South-East Asia and the Mediterranean. They frequently contain antiseptic and body-cooling ingredients like chillies, and they stimulate digestion. They also coat the foods with which they are served – often noodles of one kind or another – quite deliciously. Flavoured oils and vinegars make for wonderful salad dressings.

Of the two preserving mediums, vinegar is the more effective as its acidity is anathema to most bacteria. Infused vinegars therefore tend to keep longer than their oil equivalents. Oils can go rancid relatively quickly once exposed to the air, although the ingredients used to flavour them often have antibacterial properties. They are also not immune to the *Botulinum* bacterium. Their lives can be extended by heating them under pressure (see Bottling and Canning chapter), but this may impair the taste. The best compromise is to keep them in the fridge and to monitor them closely for spoilage. Once they are opened, you will need to use them within a couple of weeks. Fortunately, oil-based pastes like pesto freeze very well.

Herbs are not the only foods that can be used as the basis for feisty infusions and pastes. As you will see from the recipes that follow, certain vegetables and fungi can perform similar tricks.

THAI RED AND GREEN CURRY PASTE

In the UK you can now find Thai food in out-of-the-way rustic pubs where you half expect them to still refer to the place as Siam. Thai cooking is taking over the world and red and green curries are the biggest weapons in its armoury.

The difference between the two kinds of paste is largely cosmetic. The red one takes its colour from dried red chillies, the green from fresh green ones. As the Heinz marketing people have found with their ketchup, people are capricious – sometimes they feel like eating red, at other moments they fancy a bit of green. In Thailand the red version is usually the hotter, whereas in some European Thai restaurants the 'natural' semiology is demonically reversed. If you order a green curry in France on the assumption it will be mild, you may get a similar shock as the unwary traveller who drinks from the tap marked 'C' (in other words the hot one).

THAI CURRIES

Both our pastes are similar in strength. All you need to do with either to make an authentic Thai meal is combine it with fresh chicken stock, coconut milk and shredded chicken or seafood. But unlike the ready-made pastes available in oriental grocers, ours are heat-treated to give them long lives. They will keep in the fridge for at least 6 months.

RED CURRY PASTE (MAKES 500G/1LB 2OZ)

Zest of 2 limes
1 medium head of garlic, cloves peeled
75g (3oz) whole lemon grass, roughly chopped
15 dried medium-strength red chillies, stalks removed and soaked in warm water for 20 minutes
6 fresh bird's eye chillies
100g (3½oz) galangal, peeled and roughly chopped
75g (3oz) coriander roots with a little stalk, washed thoroughly
100g (3½oz) shallots, peeled and roughly chopped
1 tablespoon shrimp powder or 1 teaspoon belachan (fermented shrimp paste)
1 level tablespoon ground cumin
1 teaspoon freshly ground black pepper
2 tablespoons sesame oil
4 tablespoons vegetable oil

Place the lime zest, garlic, lemon grass, reconstituted dry chillies, fresh chillies, galangal, coriander roots, shallots and shrimp powder or belachan in a food-processor and blend until smoothish. Fill into jars (see page 164) and refrigerate.

Put the mixture in a saucepan and add the cumin, pepper and oils. Rapidly heat, stirring constantly. As soon as the paste starts to boil, transfer to a sterilised container with an airtight lid (see page 164). Cool in a bowl of iced water, then store in the fridge.

GREEN CURRY PASTE (MAKES 500G/1LB 2OZ)

Zest and juice of 2 limes
75g (3oz) galangal, peeled and roughly chopped
14 small hot green chillies, stalks removed
1 medium head of garlic, cloves peeled
75g (3oz) roughly chopped whole lemon grass
75g (3oz) fresh coriander roots with a little stalk, washed thoroughly
100g (3½oz) shallots, peeled and roughly chopped
1 tablespoon shrimp powder or belachan (fermented shrimp paste)
1 level tablespoon ground cumin
2 tablespoons sesame oil
3 tablespoons vegetable oil
1 teaspoon freshly ground black pepper

Blend the lime zest, galangal, chillies, garlic, lemon grass, coriander roots, shallots and shrimp powder or belachan until fairly smooth.

Transfer to a saucepan, then add the cumin, sesame oil, veg oil, black pepper and lime juice. Rapidly heat, stirring constantly. As soon as the paste starts to boil, transfer to a sterilised container with an airtight lid (see page 164). Cool in a bowl of iced water, then store in the fridge.

THAI RED CURRY WITH SALT COD

Despite all the good historical explanations, the British obsession with cod remains something of a mystery to us. It isn't the most prepossessing fish and the taste is unremarkable, yet this dull beast has been hunted to the verge of extinction. In our view, cod is actually improved by curing and drying. The flesh becomes meatier and works very well at the heart of a Thai red curry. **Serves 4**

250g (9oz) salt cod (see page 48)
½ tablespoon sesame oil
1 tablespoon vegetable oil
2 cloves of garlic, chopped
1 tablespoon Thai red curry paste (see page 123)
400ml (14fl oz) coconut cream
600ml (1 pint) fish or chicken stock

10 baby corn, sliced
2 stalks lemon grass, sliced
6 kaffir lime leaves
150g (5oz) mangetout
1 small bunch Thai basil, chopped
2 tablespoons fish sauce
1 tablespoon soy sauce
3 spring onions, chopped

- Soak the salt cod in fresh water in the fridge for 24–48 hours, changing the water at least once, then drain and roughly chop.
- Heat the sesame and vegetable oils in a large pan over moderate heat. Add the garlic and cod. Fry for 5 minutes until the fish starts to colour.
- Add the curry paste, stir it in and fry for a couple of minutes.
- Pour in the coconut cream and stock, add the baby corn, lemon grass and lime leaves and simmer for 10 minutes.
- Finally, add the mange tout, basil, fish sauce, soy sauce and spring onions.
- Serve immediately with rice or noodles.

RENDANG CURRY PASTE

Another great South-East Asian curry, rendang is based on a hottish, aromatic paste of the same name. It comes from the Padang region of Western Sumatra, whose inhabitants use it to help tenderise the tough local buffalo meat.

MAKES APPROX 800G (1¾LB)

- 25g (1oz) medium dried red chillies, stalks removed (in Malaysia they often use smoke-dried chillies)
- 125g (4½oz) medium-strength, fresh red chillies, roughly chopped
- 80g (3¼oz) whole lemon grass, roughly chopped
- 60g (2½oz) belachan (fermented shrimp paste)
- 50g (2oz) garlic, peeled
- 50g (2oz) fresh kunyit or 25g (1oz) powdered turmeric*
- 90g (3¼oz) fresh galangal, peeled and roughly chopped
- 180g (6½oz) shallots, peeled and roughly chopped
- 300ml (½ pint) vegetable oil
- 4g (⅛oz) salt

*Kunyit is essentially fresh turmeric root. It looks like yellow ginger and stains the fingers yellow. If you can't get hold of it, use powdered turmeric, but only half as much.

Re-hydrate the dried chillies in hot water for 20 minutes.

Put them in a food-processor along with the fresh chillies, lemon grass, belachan, garlic, kunyit/turmeric, galangal and shallots. Blend until grainy. If you find that the paste is sticking to the sides of the blender, add the oil.

Pour into a saucepan along with the salt and gently heat. Simmer for 15 minutes, stirring constantly. Pour immediately into sterilised containers (see page 164) and keep in the fridge. The paste will keep for at least 3 months or 12 if frozen.

RENDANG CURRY

Making a traditional rendang involves slow-cooking the meat for hours while the coconut cream in which it simmers reduces to a thick paste. Rendang is usually made with beef or chicken. Fortunately for Johnny, Nick has a formidable Malaysian wife who makes him cook such delicacies. Nick has devised the following recipe for a quick version. Serves 2

- 300g (10½oz) rump steak
- 1½ tablespoons sugar
- 1 tablespoon vegetable oil
- 1 tablespoon rendang curry paste (see left)
- 1 tablespoon fish sauce
- 200ml (7fl oz) coconut cream
- 1 spring onion, chopped

- Take the rump steak and slice thinly.
- Fry over medium heat with 1 tablespoon of the sugar and the vegetable oil.
- Continue to cook until the sugar starts to caramelise, then add the rendang paste, fish sauce, coconut cream, the remaining sugar and the spring onion. Simmer for 5 minutes, then serve with noodles and sugar-snap peas.

PESTO

The medieval Genoans are said to have created pesto in imitation of the crushed walnut sauces they sampled while operating trading posts in the Black Sea. Back in the Mediterranean, they replaced the walnuts with pine nuts and found they had the perfect lubricant for their pasta.

The essence of pesto is its simplicity. Why then is the shop-bought kind so often disappointing? Nick knows the answer, having been to the factories where they make it. Commercially produced pestos are based on dried or frozen basil, hence their lack of character. You are much better off making your own. In the UK, September is the time to do it, when your home-grown basil is bushy and bursting with fragrant oils.

100g (3½oz) fresh basil, stalks removed
Juice of 2 medium lemons
100g (3½oz) Italian pine nuts, dry-roasted in a frying
  pan until lightly coloured
¼ teaspoon salt
4 cloves of garlic, peeled
150ml (¼ pint) olive oil
100g (3½oz) finely grated parmesan
Freshly ground black pepper

Blend the basil, lemon juice, pine nuts, salt, garlic and olive oil into a mixture resembling the texture of couscous. Transfer to a bowl using a spatula, then stir in the parmesan and black pepper. Easy or what?

The pesto can be stored in a sterilised airtight container (see page 164) in the fridge for up to 2 weeks, or you can freeze it for up to 6 months.

If you are feeling adventurous, you could try replacing the basil with another herb such as coriander or parsley. The pine nuts are also negotiable; toasted almonds or walnuts could be substituted.

Pesto is particularly good with penne as it sticks to their convoluted surfaces. Try it sprinkled with a few crispy cubes of pancetta (see page 70).

WARM CHICKEN PASTA SALAD WITH PESTO

This versatile salad can be eaten warm or cold, as a starter or in a picnic. It is very easy to make, hence the shortness of the list of ingredients. The pesto here is slightly different from the one we showed you how to make on the opposite page in that it contains roasted rather than raw garlic. This gives it a mellower, sweeter taste. Making it is a perfect way to use up your summer herbs before the frost kills them. **Serves 4**

THE PESTO

As per recipe on page 126, but replace the raw garlic cloves with half the pulp from a roasted head of garlic

Dribble of olive oil

THE SALAD

2 large chicken breasts, cut into long strips, seasoned with salt, pepper and lemon juice

1 tablespoon olive oil

75g (3oz) wild rocket, washed

200g (7oz) dried penne pasta, cooked, then coated in a little olive oil

A few toasted pine nuts

Cracked black pepper

- To roast the garlic, tear off a sheet of baking foil and place an entire head on it with the top sliced off. Pour a dribble of olive oil into the garlic and wrap the foil around it. Roast at 220°C/425°F/gas mark 7 for 30 minutes or until soft. Leave to cool, then squeeze out the sweet pulp. Add half of it to the other pesto ingredients and blend. You'll be left with more than you need for this recipe, so store the excess in the fridge.
- Fry the chicken breasts in 1 tablespoon of olive oil over moderate heat until nicely browned, turning them over as you go. This should take about 5 minutes.
- Mix together all the salad ingredients in a large bowl with 2 heaped tablespoons of pesto.
- Season with a little cracked black pepper.

SUN-DRIED TOMATO PASTE

Sun-dried tomato paste is tastier than the regular purée and every bit as versatile. It can be used in much the same way, for instance, as a pizza topping or an ingredient in Bolognese sauce, but also has applications all its own. Try it in a sandwich with Gruyère cheese and sweet-cured ham.

To make the paste, you can either use genuine sun-dried tomatoes or 'fully dried' ones as per our instructions on page 28.

100g (3½oz) sun-dried tomatoes
1 medium head of garlic, roasted
150ml (¼ pint) olive oil
½ teaspoon salt
½ teaspoon dried oregano

Place the sun-dried tomatoes in a small pan and just cover them with water. Simmer for 10 minutes, then leave them to soak for 20 minutes more. Pat them dry with kitchen paper.

Roast the garlic according to the instructions on page 127.

Blend all the ingredients together until smooth. Store the paste in the fridge in a sterilised airtight container (see page 164) for up to 3 months.

'SUPER' HARISSA PASTE

Like all the pastes in this chapter, harissa captures the distinctive flavours of its homeland in a highly convenient form. In this case, the region of origin is North Africa. The harissa sold in the local souks is a fiery, pared-down version which needs rounding out at home with additional herbs and seasonings. Nick has simplified matters by adding the extras at the paste stage.

Simply fry up some lamb with onions and a dollop of this paste, add tomatoes and stock and you have a gourmet meal on your hands. Serve with fluffy couscous.

50g (2oz) medium-strength dried cayenne chillies*
100ml (3½fl oz) olive oil
5 cloves of garlic, peeled
1 tablespoon ground cumin
1 tablespoon ground coriander
1 teaspoon caraway seeds
Juice of 1 lemon
25g (1oz) fresh mint leaves
25g (1oz) fresh parsley leaves
25g (1oz) fresh coriander leaves
2 teaspoons salt

* You could also use guajillo chillies which have a fiery liquorice flavour.

Remove the stalks from the dried chillies, then soak them in hot water for 30 minutes. Blend them with the remaining ingredients until you have a smooth paste.

Store in the fridge in a sterilised airtight container (see page 164) for up to 1 month, or in the freezer for up to 6 months.

TAPENADE

Tapenade is a pungent, salty speciality of the French Riviera (*tapeno* is the Provençal word for caper). There is something ancient about the flavour, yet amazingly, tapenade was invented in the late nineteenth century by a chef at the Maison Dorée in Marseilles.

This recipe is more substantial than some versions as it incorporates chopped tuna. Try it with seared fish steaks, a roasted vegetable salad or stuffed into tomatoes. Alternatively, serve the tapenade as a dip with pitta bread.

> 200g (7oz) black olives, weighed after the stones have been removed
> 10 anchovy fillets, rinsed first if preserved in salt (see page 45)
> 50g (2oz) capers, rinsed before use if preserved in salt (see page 52)
> 2 cloves of garlic, chopped
> 25g (1oz) fresh basil
> About 3 sprigs of fresh thyme
> 75g (3oz) tuna tinned in brine, drained
> 1 teaspoon Dijon mustard
> 150ml (¼ pint) olive oil
> 30ml (1fl oz) lemon juice
> Ground black pepper

Place the olives, anchovies, capers, garlic, basil and thyme in a food-processor and blend until grainy in texture.

Chop the tuna quite finely and add it to the blended mix with the mustard, olive oil, lemon juice and black pepper.

Mix thoroughly with a wooden spoon and put in a sterilised jar/pot (see page 164). It will keep for up to 1 month in the fridge.

SUN-DRIED TOMATO SOUP WITH TAPENADE

It isn't easy to make a really good tomato soup, but we think we may have achieved it here. We were thinking of selling the recipe to Heinz, but we've decided to give it to you instead. Serves 4

> 100g (3½oz) carrots, sliced
> 100g (3½oz) celery, sliced
> 150g (5oz) onions, sliced
> 25g (1oz) butter
> 4 teaspoons olive oil
> 1 litre (1¾ pints) vegetable stock
> 60g (2½oz) plain flour
> 150g (5oz) tomato purée
> 40g (1½oz) sun-dried tomatoes (see page 28), simmered in water for 10 minutes until soft
>
> A pinch of white pepper
> 1 teaspoon salt
> A pinch of allspice
> A pinch of ground bay leaf
> 10ml (⅓ fl oz) cider vinegar
> 40g (1½oz) sugar
> 25 basil leaves
> 150ml (¼ pint) double cream
> 100g (3½oz) tapenade (see left)

- Fry the carrot, celery and onion in the butter and olive oil over moderate heat until soft.
- Heat the vegetable stock in a separate pan.
- Add the flour and the tomato purée to the fried vegetables and stir in well. Then turn the heat down and start adding the hot vegetable stock. Do this gradually, stirring constantly to make a smooth soup.
- When all the stock is incorporated, add the sun-dried tomatoes, pepper, salt, allspice, ground bay leaf, vinegar and sugar and simmer for 15 minutes.
- Add the basil and blend until smooth. Then stir in the double cream.
- Serve in soup plates. Garnish with tapenade and have hunks of baguette ready on the side.

ADOBO

If you've followed our advice on page 27 and smoke-dried some jalapeño chillies to turn them into chipotles, this is what to do with them next. Although the Filipinos confusingly use the term 'adobo' for a stew made with soy sauce and garlic, for us it will always remain this delectable Mexican vinegar paste.

Adobo works equally well as a marinade or a sauce base and it gives a definite lift to items destined for the barbecue. Use it as a marinade for beef and pork.

You don't have to stick to chipotles when making adobo, even within a single batch. Winning alternatives include the pasilla or 'raisin chilli', shiny, curved and mildly spicy, and the gently hot ancho, the mature, red, heart-shaped form of the poblano chilli. If you haven't got round to growing your own, Mexican chillies are readily available through mail-order companies and specialist shops.

MAKES APPROX 350G (13OZ) OF PASTE, ENOUGH FOR 2 MEXICAN FEASTS

- 1 medium head of garlic
- 1 teaspoon olive oil
- 2 chipotle chillies (see page 27), slit open and seeds removed
- 5 ancho chillies, slit open and seeds removed
- 1 teaspoon cumin seeds and 1 teaspoon coriander seeds, dry-fried in a pan until lightly coloured
- 25g (1oz) fresh oregano
- 2 shallots, peeled and roughly chopped
- 1 teaspoon ground cinnamon
- 2 teaspoons salt
- 125ml (4fl oz) red wine vinegar
- 50ml (2fl oz) balsamic vinegar

To roast the garlic, slice the top off the head to expose the flesh and dribble on the olive oil. Wrap the garlic in foil and bake in a hot oven (220°C/425°F/gas mark 7) oven for 25–30 minutes or until soft. Allow to cool, then squeeze the pulpy flesh out into a bowl.

Blend the chipotle and ancho chillies with the cumin and coriander seeds until smooth. Add the oregano, garlic and shallots and continue to blend until they have been incorporated.

Mix in the rest of the ingredients and stir into a paste.

Store in the fridge for up to 6 months in a sterilised airtight container (see page 164).

BARBECUED CHICKEN IN ADOBO

Adobo is made for summer. Its spicy oiliness makes it an excellent marinade for chicken, particularly when it's heading for the barbecue or char-grill. Serve this up with a black bean salsa and you have a satisfying and authentic Mexican meal. **Serves 4**

4 chicken legs

75g (3oz) adobo paste
  (see left)

THE SALSA

125g (4½oz) black beans,
  soaked overnight in water

1 red jalapeño chilli, seeds
  removed and finely
  chopped

1 green jalapeño chilli,
  seeds removed and finely
  chopped

1 large red onion, chopped

Juice and zest of 1 lime

1 tablespoon olive oil

1 small bunch of oregano,
  chopped

1 small bunch of coriander,
  chopped

1 teaspoon honey

A splash of red wine
  vinegar

- Marinate the chicken legs in the adobo paste for at least 4 hours prior to cooking.

- Drain, then boil the soaked black beans in enough water to comfortably cover them and simmer until soft. This will take about 45 minutes. Drain the beans and refresh with cold water.

- Place the drained beans in a large bowl with the chillies, onion, lime juice and zest, olive oil, oregano, coriander, honey and vinegar. Stir thoroughly and store in the fridge until the chicken is ready.

- Cook the chicken over a barbecue until somewhat charred on each side. It must be cooked right through, underdone chicken being a no-no from the perspective of both hygiene and taste.

- Spoon a large portion of salsa onto each plate and serve with the chicken. If you also provide soft flour tortillas, your guests will be able to roll their own enchiladas.

TRUFFLE BUTTER

If you ever find yourself in possession of a fresh white 'Alba' truffle, we'd frankly advise you to use it immediately. Its perfume, initially almost indescribably intense, will fade by the hour. This is one of the reasons truffles are so expensive – we're talking several pounds a gram. But a little goes a long way, and if you find yourself with any left over, you could do a lot worse than transform it into this butter. Nick learned this the hard way. He made it his life's work to harness the powerful but transient flavour of *Tuber magnatum pico*. Every November he'd spend all his Christmas money on a few small nuggets of white truffle. He'd carry them around in his pocket, burbling to strangers about how he was going to turn them into gold, and overstimulating every goat and pig within miles. Then he'd go home and turn his kitchen into a laboratory.

He tried heating truffles to exactly 68°C/154°F. He soaked them in alcohol and froze them in oil and everything in between. All in vain. In the end, his truffle dealer put him out of his misery. This is our friend Mike de Stroumillo. Nick's obsession had been a nice little sideline for him, but news of yet another hundred quid's worth going up the chimney was too much for his conscience. 'Look, have you tried preserving them in butter?' he said with a sigh. So Nick did. He served the truffle butter with artichoke hearts and home-made pancetta and his hairy brother-in-law came over all amorous.

WHITE TRUFFLE BUTTER IS THIS SIMPLE TO MAKE
10g (½oz) fresh white truffle, finely chopped
250g (9oz) unsalted butter, at room temperature

Thoroughly mix the truffle into the butter. Portion into an ice tray, then store in the fridge for up to a week or in the freezer for up to 3 months.

TRUFFLE OIL
You can also flavour oil with white truffle (see photo opposite). To do this, immerse ultra-thin slices of the fungus in a neutrally flavoured oil such as rapeseed. Do this at the rate of 5 slices per 100ml (3½fl oz). Unfortunately, the taste will deteriorate after a few days whatever you do. Such is the way of the elusive truffle…

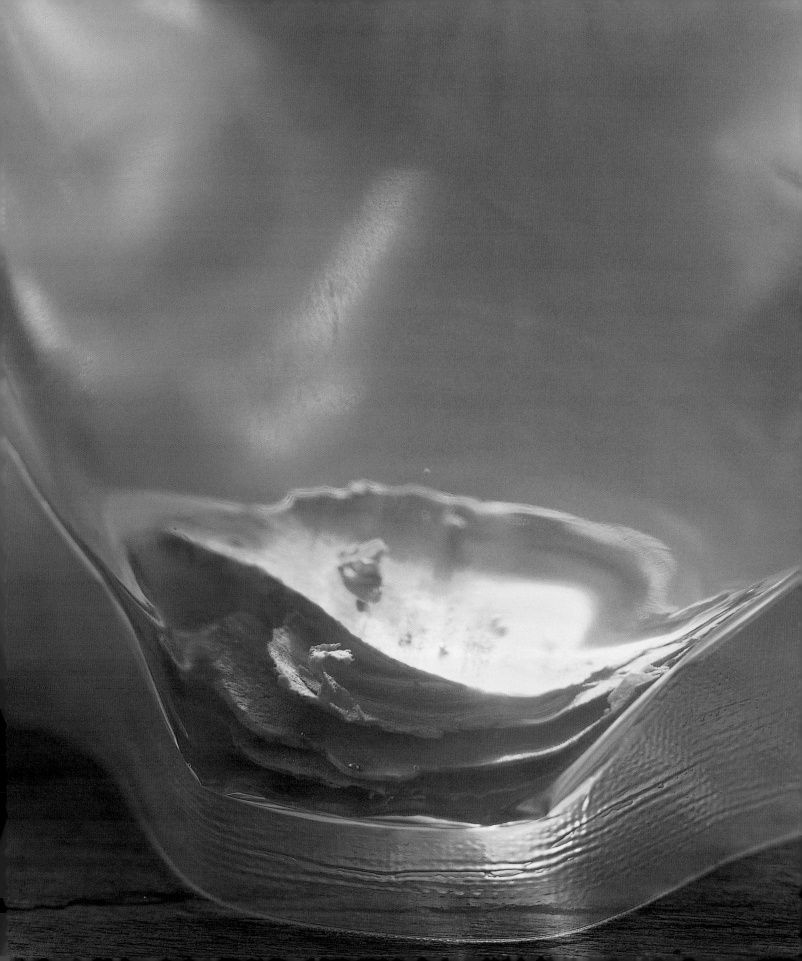

CHILLI OIL

Chilli oil is a luxurious way of adding fire to a dish. You can drizzle it on a pizza, incorporate it into a salad dressing or use it as a cooking medium for a punchy stir-fry. One of the beauties of chilli oil is that its flavour stands up well to high temperatures.

There are a number of ways to make it. The simplest is just to immerse whole chillies in oil, puncturing them with a pin if you want them to sink. A slightly more sophisticated way of making chilli oil is the following:

500g (1lb 2oz) mild red chillies, seeded
250ml (9fl oz) olive or other neutral oil, e.g. corn

Run the chillies through a juice extractor, then reduce the juice in a saucepan until it darkens and starts to thicken. You are aiming to reduce its volume by approximately 90 per cent.

Add the thickened chilli juice to the oil and funnel into a sterilised bottle (see page 164). It will keep in the fridge for up to 1 month.

HORSERADISH OIL

Horses are lucky in our opinion. Their radishes are are lot more interesting than 'human' ones. This oil packs quite a kick and is surprisingly versatile. Naturally it is excellent drizzled on roast beef, but Nick enjoys it with cheese on toast. You could even use it as an alternative to wasabi mustard, mixing it with soy sauce to make a dip for sushi or sashimi.

 250ml (9fl oz) oil*
 1 small horseradish root, peeled, then thinly sliced
   or cut into juliennes

*The best oils to use are neutral in taste, such as corn or sunflower oil. You could, however, mix in a dash of toasted sesame oil to give the finished product an appealing nuttiness.

Mix the oil with the horseradish, then store in a sterilised airtight container (see page 164) in the fridge for up to 2 months.

LEMON OIL

The most obvious way to use this oil is as the basis for a lemony salad dressing, but as usual your imagination is the only limit. The flavour combines very well with fish and lamb.

 Zest of 4 lemons
 500ml (18fl oz) olive oil or neutral oil such as
   sunflower oil

Combine the lemon zest with the oil, then store in a sterilised airtight container (see page 164) in the fridge for up to 6 months.

ROASTED GARLIC OIL

This is a great way of preserving roast garlic for dressings, sauces and marinades. The ratio of garlic to oil is quite high, so this is more of a paste than an infusion. It is nice and sweet and very convenient.

 2 medium heads of garlic, roasted
 200ml (7fl oz) olive oil, plus 2 teaspoons

To roast the garlic, slice the tops off the heads to expose the flesh and dribble 1 teaspoon of olive oil onto each. Wrap them in foil and bake in a hot oven (220°C/425°F/ gas mark 7) for 25–30 minutes or until soft. Leave to cool, then squeeze out the pulpy flesh into a bowl.

Combine the garlic with the remaining olive oil and store in the fridge in a sterilised airtight container (see page 164) for up to 3 months.

This paste is excellent for basting roast meat, for instance, chicken. Just add lemon juice and a little tarragon. On table-tennis nights, we're particularly fond of slow-roast pork belly with rosemary, sage and roasted garlic oil.

FENNEL OIL

There's something thought-provoking about the affinity between fish and fennel. The two species have precisely nothing to do with each other while they are alive, yet they taste as though they were made for each other.

Using fennel to flavour oil is a good way of capturing its liquoricy essence. Serve it with poached cod or salmon.

2 heaped tablespoons fennel seeds
400ml (14fl oz) light olive oil

Blend the fennel seeds in a spice mill or food-processor, then combine with 1 tablespoon of warm water and stir into a paste.

Mix the paste into the oil, then leave for a few days, stirring occasionally.

Carefully pour the oil through muslin into a sterilised airtight container (see page 164), leaving the sediment behind.

Store in the fridge for up to 6 months.

COD IN FENNEL OIL

'You must be mad!' shouted Johnny. 'You can't deep-fry cod without batter, especially not in fennel oil!' 'I am NOT deep-frying the cod! I'm slowly poaching it,' Nick explained with a sigh. 'This oil is the perfect cooking medium for chunky cod fillets like this one. Using it allows the aniseed aroma of the fennel to permeate deep into the flaky flesh.' Then he fed Johnny a forkful to shut him up. **Serves 2**

2 chunky cod fillets
Some cooking string
Fennel oil (see left) –
** enough to cover the fish**
** in the pan, probably**
** around 350ml (12fl oz)**

Salt
Cracked black pepper
A splash of lemon juice

- Tie up the cod fillets as if you were wrapping parcels, bunching the meat up until they are vaguely box-shaped. The string should be firm but not tight. Its function is to hold the fish together during the poaching process.
- Pour the fennel oil into a non-stick frying pan, then lower in the fish fillets.
- Gently heat the oil to somewhere between 80–90°C (176–194°F). A food thermometer will prove more than handy here. Simmer the fish at this temperature for 15 minutes until cooked through.
- Before you gently lift the fillets out of the pan, make sure that the accompaniments are ready. We recommend serving the cod with mashed potato and fine green beans.
- Lay the fish on the potato and season with salt, pepper and a squeeze of lemon juice. It is great with aioli, a garlicky mayonnaise made with olive oil.

RASPBERRY VINEGAR

Raspberry vinegar (see photo on page 120) is often used to make pleasantly sweet salad dressings. You can also melt it with butter and pour onto grilled or fried fish. Alternatively, use it with a dash of port to de-glaze a roasting tin ahead of making gravy.

900g (2lb) fresh or frozen raspberries
600ml (1 pint) red wine vinegar

Place half the raspberries in a bowl and cover with the vinegar. Cover with a cloth and leave for 5–7 days in a warm place, stirring occasionally. Then strain the liquid, pour it over the rest of the raspberries and cover and leave as before. Strain into sterilised bottles (see page 164), adding a few whole berries for visual effect. Wait at least 1 month before using.

TARRAGON VINEGAR

Tarragon goes remarkably well with fish and chicken, so tarragon vinegar is ideal as a basis for dressings to accompany said beasts. It also looks good and can be used to make classy tartare, bearnaise and hollandaise sauces.

25g (1oz) fresh tarragon
600ml (1 pint) white wine vinegar

Bruise the tarragon to release its essential oils, pack it into a jar and pour over the vinegar. Seal and shake thoroughly. Leave in a warm place for 2–3 weeks before using, shaking every day.

Strain through muslin into a sterilised bottle and seal (see page 164). Store in a dark cupboard for up to 1 year.

SALMON WITH TARRAGON VINEGAR HOLLANDAISE

In the absence of salmon, this hollandaise would be equally delicious with sea or rainbow trout. This version is unusually stable for such a notoriously fickle sauce: you can store it in the fridge for up to a week. It's also much easier to make than cooking lore would suggest. **Serves 4**

THE HOLLANDAISE
175ml (6fl oz) milk
125ml (4fl oz) cream
12.5g (½oz) corn flour
A pinch of cayenne pepper
A pinch of white pepper
½ teaspoon salt
30ml (1fl oz) lemon juice
125g (4½oz) egg yolks
   (roughly 5 yolks)
1 tablespoon tarragon
   vinegar
40g (1½oz) butter

THE SALMON
4 salmon fillets, skinned or
   scaled and scored,
   seasoned with a little salt,
   black pepper and a
   dusting of flour
1 tablespoon butter,
   clarified if you're feeling
   pernickety, plus an extra
   knob of butter
Juice of ½ a lemon
Some sprigs of chopped
   tarragon

- Pour the milk and cream into a saucepan with the corn flour, cayenne pepper, white pepper, salt and lemon juice. Whisk the mixture constantly as you slowly heat it. Once the corn flour has thickened, cook for a further minute, then remove the pan from heat.
- Immediately whisk in the egg yolks and the vinegar, then briefly return the pan to the heat (for less than a minute).
- Stir in the butter and the hollandaise is ready to serve. Store any excess in an airtight container in the fridge; it will keep for up to 1 week.
- Using a non-stick pan, fry the salmon in 1 tablespoon of butter over moderate heat for around 4 minutes on each side. Remove the fillets from the pan and place them on plates.
- Heat up the extra butter in the same pan until it has boiled off but hasn't started to burn, then squeeze in the lemon juice. The butter will fizz up. Pour it over the salmon.
- Serve with hollandaise sauce and garnish with tarragon.

FERMENTING

Most of the techniques in this book are aimed at the suppression of micro-organisms. Fermenting is different. In the fifth century BCE, the Greek author Herodotus captured the essential strangeness of this approach when he wrote in amazement, 'All men are afraid of their food spoiling, but the Egyptians make dough which has to be spoiled.'

He was referring to the miracle of leavening. The locals had learned that yeast, which formed spontaneously on crushed dates and figs, somehow dramatically inflated their breads.

As with other preserving methods, most of the benefits of fermentation were probably discovered by accident. The Koreans have a legend in which a poor farmer puts some withered old cabbages in a bowl of sea water to revivify them. When he looks at them a couple of hours later, he is overjoyed to discover they have swollen magnificently. This is matched by his disappointment the next day when he returns to find them limp and even smaller than they were to begin with. But there is a consolation: they taste delicious. A similar revelation occurred when a forgotten tub of fruit turned out to have decomposed into a juice which was good to drink and caused a pleasant intoxication. Or when a bird left hanging for a few days proved tastier and more tender than ones eaten immediately.

The fermented specialities of some parts of the world may smack of desperation to cosseted urban Westerners, but the people who depended on them grew to adore them. A case in point is Norwegian *rakfisk*, traditionally made by burying tubs of oily fish in the soil, where they ferment in their own enzymes. The end result is overpowering to the unaccustomed. The same can be said of Sudanese *um tibay*. This is made by finely chopping an entire gazelle (hooves, bones, offal, nerves, the lot), stuffing the mixture into the animal's stomach and leaving it to hang on a tree for a few days where it ferments in the super-hot sun. Then it is cooked in hot ashes, cut into strips and dried. Nothing is allowed to go to waste.

Fermentation has an astonishing range of practical applications, from compost-making to the manufacture of synthetic rubber. In the field of food and drink, it is used to make so many kinds of product that a line must be drawn somewhere or we could go on forever. Accordingly, in this chapter we are going to ignore brewing, baking and the fermentation of dairy products. But just imagine how different the course of history would have been had people never developed these ways of exploiting microbial activity.

In the meantime, there are plenty of excellent fermented products you can make at home. We've already shown you how to make two meat ones: paprika salami and garlic sausage (see pages 90 and 93). Here we teach you to prepare four more classic fermented foods. Cumulatively they illustrate what fermenting is all about: harnessing natural processes to produce diverse and amazing and transformations. If you find yourself addicted to this rewarding form of alchemy, there are endless avenues to explore.

THE PRINCIPLES OF FERMENTING

The basic idea in fermenting is to promote the growth of certain bacteria, moulds or yeasts which produce something useful as a by-product of their life-sustaining activities. This varies from carbon dioxide (the key to making bread rise) to alcohol (obviously handy when it comes to brewing) to lactic acid (the secret behind cheese-making and the long lives of cured sausages). Many of these substances in turn inhibit the growth of unwanted beasties. They also give fermented foods their characteristic flavours. To get the balance right, however, you need to be precise in your preparations.

The first requirement, as ever, is hygiene. All tools and vessels used in fermenting must be properly sterilised to prevent the end-products from spoiling. The next imperative is to give the benign organisms something to eat. One way or another, this is usually sugar. Fortunately, this tends to be present in the food you are trying to preserve, for example, as glucose and fructose in vegetables and lactose in milk products.

The final part of the jigsaw is storage. Conditions must be right for the 'good' organisms to proliferate, thereby swamping potential undesirables. The food must be properly sealed, and the pH (acidity) of its environment may need to be adjusted, often with vinegar. Salt may be added to inhibit the growth of 'bad'

bacteria. Then it is primarily a matter of keeping the food at the right temperature. Too cold and the benign organisms will reproduce too slowly. Too hot and they may go bananas. If this happens, the end-products will be unpalatable at best and at worst downright dangerous. Fortunately, the tell-tale signs are hard to miss. These include alarming discolouration, abundant mould growth and repugnant aromas (although sauerkraut and kimchee don't smell great at the best of times).

For the majority of fermenting processes, a temperature of between 15–20°C (59–68°F) is ideal. We'll give you full instructions for the preparation of the items covered in this chapter as we come to them.

SAUERKRAUT

Although it is now most closely linked with Germany, fermented sliced cabbage has been a part of life in the Far East for thousands of years. The strange truth is that the men who built the Great Wall of China subsisted largely on sauerkraut. The chief difference is that theirs was fermented in rice wine, whereas the German kind relies on salt. Fermented cabbage appears to have been brought to the West by migrating Central Asian tribes and the armies of Genghis Khan.

Sauerkraut proved immensely popular in central and northern Europe, and later in the colder parts of North America. Cabbages were easy to grow, preserving them was straight-forward and the end-product was rich in vitamin C. Sauerkraut was particularly valuable in the winter when fresh fruits and vegetables were unavailable. The British Navy equipped its ships with barrels of it to ward off scurvy.

As with many popular fermented foods, the chemical work underlying sauerkraut is done by lactobacillus bacteria. These microbes convert the natural sugars in the cabbage to lactic acid, helping to preserve it and giving the finished product its characteristic sour tang.

There is no point pretending sauerkraut smells great as it's cooking, but the finished item is lip-smackingly tasty. Try it with home-made frankfurters (page 97).

TO MAKE SAUERKRAUT
Two heads of cabbage will yield quite enough sauerkraut for your immediate needs:

> 1 very large container, preferably glass or enamel,
>   but plastic will do
> 2 heads of cabbage
> Sea salt

Clean and sterilise the container by pouring boiling water down the sides and letting it sit for a few minutes.

Remove the core from the cabbages, then cut them into quarters and finely shred them. Food-processors do this well.

Weigh the cabbage and add 2 heaped tablespoons of salt per kilo (2¼lb). Mix them together thoroughly in the fermenting container. The salt will soon start to draw the juices from the cabbage, which will ferment in its own brine.

To make sure the cabbage stay immersed in it, weigh them down with plastic food bags filled with brine made with 1¼ tablespoons salt per litre (1¾ pints) of water. If the bags leak a little, the cabbage won't be watered down. Alternatively, use a plate weighed down with a jar of water, as in the photo opposite.

Cover the container with clingfilm or wrap it in a heavy cloth. Then leave it in a dark place to ferment for 6 weeks at ambient temperature (18°C/64°F).

When the the sauerkraut is ready, decant it into sterilised jars, seal (see page 164) and store them in the fridge for up to 6 months. Alternatively, you can bottle it (see chapter 10) in which case it will keep for 1 year. To do this, drain off the juice, heat it to boiling point, then re-add the cabbage and simmer for 5 minutes. Transfer the sauerkraut and juice into preserving jars and process in boiling water (see page 190) for 25 minutes. Cool at room temperature and store under the stairs or in a dark larder.

BRAISED PORK WITH SAUERKRAUT

This rich Bavarian favourite is best made with a thick cut of pork belly. As you eat it, you can almost hear the slapping of palms on lederhosen-clad thighs. Sauerkraut and apple is a classic sweet-and-sour combination, and as both the Chinese and Germans have discovered, sweet-and-sour goes very well with pork. This is a warming dish for a cold winter's evening. **Serves 4**

10 sage leaves
2 Bramley apples, peeled and chopped
4 cloves
4 cloves of garlic, chopped
1 tablespoon Dijon mustard
2 tablespoons brown sugar
300g (10½oz) sauerkraut (see page 144), drained
200ml (7fl oz) chicken stock
100ml (3½fl oz) beer
A thick cut of pork belly, about 1.5kg (3lb 5oz)
Salt
Pepper

- Preheat the oven to 180°C/350°F/gas mark 4.
- Place the sage leaves, apples, cloves, garlic, mustard, brown sugar, sauerkraut, chicken stock and beer in an ovenproof dish and mix thoroughly.
- Add the pork belly, skin side up. Press the meat in firmly so that it nestles among the other ingredients. Then season with salt and pepper.
- Bake in the oven for 3 hours until the pork is soft and tender.

- Remove the pork belly and place it in a clean ovenproof dish. Finish it off under a hot grill for 10 minutes to crisp up the skin. Alternatively, turn the oven up to 220–230°C/425–450°F/gas mark 7–8 and roast the belly for a further 15 minutes.
- Shred the pork into the sauerkraut mix and serve with baked potatoes.

KIMCHEE

Kimchee is a generic term for spicy Korean fermented pickles. Johnny's wife Percy is no-one's idea of a Korean, but she did manage to get seriously addicted to kimchee. It suited her fiery nature. On one occasion, she bought a tub in Chinatown in London, took it back to her office and popped it in a drawer. Then she went off to a meeting. When she got back, the kimchee was gone. It had been escorted from the building by Security. A nervous colleague had complained about the smell.

Kimchee, particularly the cabbage variety, can indeed whiff a bit. In this, it is only obeying the law common among fermented foods of smelling one way while tasting quite another. This is what makes kimchee an acquired taste. The initial reaction is usually negative, but it is later overruled when the benefits come to light (though not in everyone). The same is true of whiskey, blue cheese and gentleman's relish. Persevere with kimchee and you may find you can't live without it.

They eat literally hundreds of kinds of kimchee in Korea. The most popular revolve around cucumber, turnips and cabbage, and are made with mind-blowing quantities of salt and hot red pepper powder. Fermented shrimp or oysters are often added to the mix. This recipe is a somewhat toned-down introduction to the genre.

TO MAKE 1 LARGE JAR

   1kg (2¼lb) pak choi or Chinese cabbage
   Salt
   3 cloves of garlic, chopped
   2 teaspoons chopped ginger
   3 medium-strength hot red chillies, finely chopped
   1 small bunch of spring onions, chopped
   2 teaspoons sugar
   2 tablespoons fish sauce
   1 tablespoon soy sauce

Separate the leaves of the pak choi or Chinese cabbage and wash them well. Then either tear or roughly chop them into small pieces.

Dissolve 3 tablespoons of salt into 750ml (1⅓ pints) of water and pour over the leaves in a large bowl. Make sure all the leaves are immersed in the brine by weighing them down with something, for instance, a plate with a jug of water on top.

Leave the pak choi/cabbage in the brine for 8 hours, then drain it off and immerse them in fresh cold water for 10 minutes. Then drain off this water too.

Now the leaves have been cured and rehydrated they are ready to ferment. Place them in a tall, sterilised glass jar (see page 164) and make up the liquor. Take 650ml (1 pint 3fl oz) water and mix in 1½ tablespoons of salt plus the garlic, ginger, chillies, spring onions, sugar, and the fish and soy sauces. Make sure the salt is completely dissolved, then pour the liquid over the leaves in the jar, taking care to cover them completely. Seal in line with the instructions on page 164.

Leave the jar in a warm room (above 24°C/75°F) for 24 hours. Then refrigerate the kimchee for up to 1 month.

Nick once made a deranged fusion dish with this kimchee. He fried some boiled diced potato in oil with onion, garlic and diced pancetta, then added potted shrimp, pak choi kimchee and some petit pois. The result was unique and startlingly delicious.

MISO

Someone coming across miso for the first time will probably have no idea what it is made from. All they will see is a yellow or brown paste. When they taste it, they will find it unlike anything they have eaten before, grainy, salty and bursting with *umami*. The Japanese have long known about this 'fifth' fundamental taste (the others are sweet, salty, sour and bitter), but only recently has science proven the existence of a tastebud receptor for it. Umami-rich foods are moreishly savoury and this is certainly true of miso.

The answer to the riddle is that miso is made from soya beans. These are boiled, then mixed with cooked rice or barley inoculated with the spore of a fungus called *Aspergillus oryzae*. Salt is added, then the mixture is packed into wooden vats and left to mature for between 6 months and 2 years. Every year, some 600,000 tonnes of miso are produced in Japan. That's almost 5 kilos (11lb) for every man, woman and child.

THE STARTER CULTURE

One component you are unlikely to have in your fridge is the aspergillus, the key to making the *koji* (starter culture). Fortunately, the same mould is used to make sake, and it is readily available from home-brewing shops and websites, Chinese pharmacies and some oriental supermarkets. You can buy it from G.E.M Cultures in California (www.gemcultures.com)

MAKING MISO

Miso comes in a range of colours and sweetnesses that reflect different soya/grain combinations. This recipe below produces a versatile light brown paste that is excellent in soups and dressings. You will end up with a lot of miso but it will keep in the fridge for a year. The beans can be bought in oriental and health food shops.

> 1.5kg (3lb 5oz) short-grain rice
> Koji starter culture (*Aspergillus oryzae*), according
>   to packet instructions
> 1.5kg (3lb 5oz) fresh soy beans
> 650g (1lb 7oz) sea salt
> A sterilised cloth – this can be done by boiling it

Wash the rice in water and leave it to soak overnight. Place it in a sieve or colander to dry for 4 hours, then steam it until cooked (this will take around 20 minutes). Cool it by tossing it in the air. For a good *koji*, the rice particles must be soft and elastic, dry on the surface but wet inside.

When the rice has cooled down to 25–30°C/77–86°F, add a small amount of the fungi spore, mixing well to distribute it. Then spread out the cloth and make small piles of rice on it, each about 10cm (4in) high. After about 10 hours, the piles will start to heat up, a sure sign that the fungus is growing. You need to control their temperature for about 40 hours, keeping it between 35–40°C/95–104°F and taking care that it doesn't rise beyond this upper limit. To do this, you adjust the size of the rice piles. If you make them them less deep, the temperature will fall and vice versa. When the incubation period is over, you may be able to see feathery white fibres of fungus on individual grains of rice.

PREPARING THE SOY BEANS

Soak the soy beans overnight, then boil them until they can be mashed easily between your thumb and little finger. This will take about 3 hours. Then drain the beans and mash them while they are still hot. When they have cooled to 35–40°C/95–104°F, mix with the *koji* rice and the salt, which should be combined together first.

Next, stuff the mixture into a keg or plastic bucket, smooth the surface and rub a teaspoonful of salt into it. Place a plastic sheet on top, then a lid cut to fit the surface area of the miso. Weigh this down to the tune of about 3kg (6¾lb). Finally, cover the container with wrapping paper and tie it up with strings.

Store the tub or bucket of developing miso in a cool, dark place. After a month, have a peek inside, but keep this brief to minimise the chances of unwanted microorganisms getting in. Check whether any liquid has formed on the surface of the miso. If it hasn't, you need to increase the weight pressing down on the lid. If and when the liquid has appeared, give the paste a mix, bringing the bottom to the top and vice versa. Reseal the container and mix again after a month. Then leave the miso in peace to mature.

After several months, the miso will start to smell good and turn yellow or light brown. You can try it after 6 months and it will keep improving for up to 18.

COOKING WITH MISO

Miso can be used to make delicious dressings, but it is most important as a basis for soups. Miso soup is a cornerstone of Japanese culture. Every household has its own take on the subject and new brides are expected to ditch the techniques they learned at home and replace them with those of their husbands' families.

Miso soup is light, rich and nutritious and very easy to make. It is also extremely versatile.

MISO SOUP WITH PORK AND NOODLES

A simple miso soup is an essential component of a traditional Japanese breakfast, but more elaborate versions can be meals in themselves. This robust rural recipe is a good example, yielding a warming and filling soup. Serves 4

1.25 litres (a generous 2 pints) dashi. You can buy instant dashi in oriental food stores or make your own (see below)

200g (7oz) dried soba noodles

1 small white radish, cut into very thin strips

125g (4½oz) carrots, cut into very thin strips

125g (4½oz) miso paste (see left)

2 tablespoons soy sauce

2 tablespoons mirin

250g (9oz) pork tenderloin, sliced very thinly (the pork will be easier to slice paper thin if you half freeze it first)

4 spring onions, chopped

- To make your own dashi, which is an important Japanese stock, take a 15cm (6in) piece of kombu (dried kelp) and 25g (1oz) of bonito flakes, both available from oriental and some 'regular' supermarkets. Add to 1.5 litres (2½ pints) of water and simmer for 20 minutes. Do not allow to come to a full boil. Allow the flakes to settle, then strain the liquid through a fine sieve and reserve. Alternatively, make up 1.25 litres (2 pints) of instant dashi according to the packet instructions
- Cook and drain the noodles according to the instructions on the packet. Refresh with cold water and reserve.
- Heat the dashi in a large pan, add the radish and carrots and simmer for 5 minutes.
- Add the miso to the soup and stir in thoroughly.
- Season the soup with the soy sauce and mirin, add the thinly sliced pork and simmer for a few minutes until cooked.
- Serve in bowls with the noodles. Garnish with spring onions.

BLACK BEAN SAUCE

With the exception of water and a little salt, black soy beans contain everything the human body needs. Their protein content rivals that of meat, but soy protein is much easier to digest. They are also about 11 per cent carbohydrate and yield the world's number one cooking oil. The soy bean, which provides tofu, miso, milk, soy sauce and a good deal besides, has justifiably been called 'the cow of China'. Here we see what happens when you ferment it.

This dark, subtly hot sauce is fantastic with fish, as Chinese take-aways have discovered from here to Timbuktu. For a richer experience, you can test your preserving patience and make it yourself. Or you can cheat a little, saving a couple of years by buying ready-fermented beans from an oriental supermarket. They are pungent and very salty, so you may want to rinse them before using them.

TO FERMENT BLACK BEANS
1 kg (2¼lb) black soy beans
300g (10½oz) salt
25g (1oz) ginger, roughly chopped
6 cloves of garlic, roughly chopped
zest of 2 oranges

Soak the beans overnight, then parboil them for about 30 minutes.

Drain them, leave them to cool, then mix them with the remaining ingredients.

Pack the mixture tightly into sterilised jars and seal according to the instructions on page 164. Store at ambient temperature (for instance, in a larder or cellar) for 1–2 years before use.

THE SAUCE – ENOUGH TO FILL ONE SMALL POT
1½ tablespoons fermented black beans (see above)
2 teaspoons finely chopped garlic
2 teaspoons finely chopped ginger
½ teaspoon finely chopped red chilli
1 tablespoon sesame oil
2 teaspoons rice vinegar
1½ teaspoons sugar
1 star anise
150ml (¼ pint) chicken or pork stock
1 tablespoon dark soy sauce
1 sprig of coriander, roughly chopped
3 spring onions, finely sliced
1 teaspoon corn flour mixed with 2 tablespoons water

Dip the black beans in a little warm water to take the salty edge off them, lifting them out with a slotted spoon.

Gently fry the garlic, ginger and chilli in the sesame oil for a couple of minutes.

Add the vinegar, sugar, star anise and black beans and simmer for a few more minutes.

Now add the chicken or pork stock and soy sauce and cook gently for 5 minutes. Then add the coriander and spring onions.

Finally, making sure that the corn flour is mixed well with the water, whisk it into the hot sauce. It should thicken immediately.

Spoon the sauce into a sterilised airtight container (see page 164). It will keep in the fridge for up to 1 week.

For a simple, satisfying dish, serve the black bean sauce with char-grilled salmon and rice.

SWEET CHILLI RIBS WITH BLACK BEAN SAUCE

The fiery quality of black bean sauce is well suited to spare ribs, particularly if you have some sweet chilli sauce to hand. This seems to douse the fire somewhat, even if technically it ought to enhance it. **Serves 4**

THE RIBS
24 pork ribs
1 litre (1¾ pints) chicken
   stock
1 small block of ginger
2 cloves of garlic
2 star anise
½ teaspoon five-spice
   powder
A dash of soy sauce

THE SWEET CHILLI SAUCE
See page 177

THE BLACK BEAN SAUCE
See page 150

- Simmer the ribs in the chicken stock, ginger, garlic, star anise, five-spice powder and soy sauce for 2 hours. If the chicken stock fails to cover the ribs, top up with water.
- Coat the ribs in a little oil and around 150ml (¼ pint) of the sweet chilli sauce. Either bake in the oven at 230°C/450°F/gas mark 8 for 10 minutes or grill on the barbecue. They won't take long as they are already cooked.
- Serve with the black bean sauce. A little goes a long way.

SUGAR

It may seem bizarre to us, in a world saturated with sugar and suffering from it, that when 'rock honey' first arrived in Europe in the early Middle Ages, it was considered an incredibly healthy substance. There were three main reasons. First, sugar fitted neatly into the prevailing system of medicine. Secondly, it had remarkable preservative powers.

120

HARD BALL
SOFT BALL

100

JAM

80

STERILISE

60

40

LIQUID
LEVEL

Made in
England

And finally, it was wildly expensive, so only a few monarchs were in danger of consuming too much. As late as the fifteenth century, by which time supply had expanded considerably, a teaspoon of sugar was worth £3 sterling in modern terms.

The feature of sugar to leave the deepest impression on the medieval mind was its ability to preserve foods with which it had been coated or impregnated. We now understand this in terms of sucrose depriving bacteria of the chance to proliferate by forming air-tight barriers and sucking up moisture. Sugar is pure carbohydrate, so although the body can use it as a source of energy, a person living on nothing else would eventually die of malnutrition, not to say thirst. Similarly, microbes cannot survive in a pure sugar environment. But all this was a mystery to contemporary alchemists and proto-scientists. Nostradamus (1503–1566) was among those fascinated by this protean substance, which assumed several distinct forms (sticky, clear, brittle, smooth, opaque, grainy, etc.) depending on the temperature to which a sugar solution was boiled and the speed with which it was cooled. It comes as a surprise to realise that the supposed great seer of the future spent much of his time writing treatises on jam.

As far as equipment is concerned, a heavy-bottomed preserving pan will prove indispensable. This will help prevent burning, the curse of many a jam-maker. The pan will be tall enough to prevent sugar solutions from boiling over, have sloping sides to facilitate evaporation, and probably feature a lip to enable easy pouring. The other essential piece of kit is a sugar thermometer, which you may already possess if you're already addicted to home-smoking. This will allow you to precisely gauge what is known in the trade as the 'height' to which you boil your sugar. This is vital if you are going to produce end-products with the desired consistency. You don't want chutneys as brittle as glass or tasting of caramel.

To illustrate the need for precision, here is the list of thirteen heights of sugar boiling identified by Frederick Bishop in 1850. They are, in ascending order of the temperature to which sugar solutions must be boiled to produce them, *petit lissé* (equating to 'smooth' in the English system), *lissé*, *petit perlé* (equating to 'thread'), *grand perlé*, *petit queue de cochon* ('little pig's tail'), *grand queue de cochon* ('big pig's tail'), *soufflé*, *petit plume*, *grand plume* (the last two are described as 'feather' by English sugar boilers), *petit boulet*, *grand boulet* (or 'ball'), *cassé* (which produces a ball which will crumble and stick to the teeth when bitten) and finally *caramel*, which snaps clean when broken.

CARAMELISED ALMOND CLUSTERS

We were faced with a dilemma here. Should we try to tackle the traditional smooth sugared almond long associated with weddings? In the end we decided not. Making such 'comfits' demands specialist equipment and lots of time. A tilting 'balancing pan' is needed to allow manipulation of the sugar syrup, which must be heated to exactly 102°C/216°F. Then the nuts are added and turned over constantly so that the syrup alternately coats them and dries to a hard shell. Several days and thousands of layers later, you have authentic sugared almonds.

It's a big ask to do all this at home, so we've decided to show you a quicker method of preserving almonds in sugar. For caramelisation, the syrup must be heated to 'crack height' (154–157°C/310–315°F), a whisker beneath its scorching point. This requires care and precision.

TO MAKE ALMOND CLUSTERS

You need two pieces of special equipment to make this treat. The first is a silicon mat. Heatproof, flexible and non-stick, these are the modern equivalents of the cold marble slabs traditionally used by confectioners to cool their wares. The next compulsory item is a thick-bottomed pan. It should have a heavy base which transfers heat evenly and responds quickly to temperature adjustments. This will give you the vital degree of control for this and several other sugar preserving processes.

600g (1¼lb) caster sugar
300g (10½oz) almonds, unskinned and unblanched

Spread the almonds on a baking sheet and heat them in a low oven (150°C/300°F/gas mark 2) until the sugar is melted and ready for them. It is important that they are warm when added to the syrup (the reason is explained below).

Pour the sugar into your pan and heat it slowly. Peer into the depths where the sugar meets the base and check that it is clear as it melts. If it is dark, you are heating it too quickly.

Stir the melting sugar gently with a metal spoon. Clumps will form, but as the minutes pass, they will become smaller and smaller until the sugar has completely melted.

At this point, take the almonds out of the oven and add them to the sugar immediately, while still hot. This is important. If the nuts were cold, the temperature of the syrup around them would drop briefly but significantly. Sugar crystals would start to form. If these precipitate in a big way in a hot syrup, you are in trouble. A chain reaction can set in, turning the whole batch grainy and leaving you with what seventeenth-century confectioners referred to as 'sugar boil'd to sugar'.

The sugar may coagulate again a bit anyway, but be patient and keep stirring. After 5 minutes or so, it will have smoothed out again. At this stage, you will hear the almonds 'cracking' as they toast in the hot syrup. The sugar should have darkened to a smoky amber. If it remains clear, turn the heat up a tiny notch.

Now pour the molten conglomerate out onto the silicon mat and leave it to cool until brittle. This will take 25 minutes or so. Break it into chunks and store it in a cool place in a biscuit tin or packed into cellophane bags. Eat within 2–3 months.

CANDIED ORANGE PEEL

Although sugar cane was probably originally cultivated in New Guinea, it was the Indians, several thousand years ago, who worked out how to isolate crystalline sugar. The Sanskrit word for the pleasing nuggets thus produced was *khanda*, hence the English term 'candy'.

To the professional confectioner, 'candy' means sugar boiled in a solution, then allowed or encouraged to recrystallise. As we mentioned in the instructions for making almond clusters, this can happen at the drop of a hat. Because recrystallisation was so easy to arrange, the earliest sweets were invariably candies.

'Candying' when applied to fruit means to soak it in syrup until most of its natural juices have been replaced by sugar via osmosis. This helps it to keep its shape and greatly increases its lifespan. In this sense, candied fruits are a kind of fossil. It's just that they have been infiltrated by a particularly appealing mineral.

TO MAKE CANDIED PEEL

Candying can transform things you wouldn't normally want to eat – in this case sour orange peel – into luxurious delights. Try dipping one end of the finished peel in molten dark chocolate and the other in granulated sugar.

One of the ingredients here is glucose syrup. This is frequently used in the confectionary trade as a 'doctor' to hinder recrystallisation. Its function here is to give the candied orange peel lustre and to prevent its surfaces from hardening.

> 1kg (2¼lb) navel orange skin* (this equates to
> 4–5kg (9–11lb) whole oranges)
> 1.8kg (4lb) caster sugar
> 200g (7oz) glucose syrup

*Try to get hold of unwaxed oranges. To skin them, cut into quarters and gently peel away the flesh. Alternatively, peel them whole and use the fruit to make Oranges in Brandy (page 184).

Cut the skin into strips and simmer them in water for about 30 minutes until soft. Drain the strips, then place them in a saucepan along with 1kg (2¼lb) of the sugar. Cover with water so that none of the peel is protruding, then simmer for half an hour. Remove the syrup from the heat and leave it to cool with the saucepan lid on.

Next day, remove the strips of peel with a slotted spoon, then add a further 200g (7oz) sugar to the syrup and heat it to boiling point, making sure that all the 'new' sugar dissolves. Remove the syrup from the heat and replace the peel. Then leave the pan to cool for another 24 hours, again with its lid on.

Repeat the process three more times, adding 200g (7oz) of sugar to the syrup on each occasion until you have used it all up. On the fifth day you add the glucose syrup instead. Bring to the boil as before, then pour over the peel.

Leave the peel covered for 24 hours, then remove it and lay it out on greaseproof paper. Allow to dry for 24–48 hours until all moisture has disappeared but the peel is still soft.

Dip in granulated sugar and store between layers of greaseproof paper in an airtight container for up to 6 months.

CANDIED ORANGE SEGMENTS

You can also candy entire orange segments in this way, with the skin still attached to the flesh. Proceed exactly as above, simmering the segments for a time before combining them with sugar. When ready, they are wonderful dipped in chocolate.

MARRONS GLACES

Chestnuts are pretty good anyway, with their melting, sweet-savoury flesh, but steeping them in syrup transforms them into genuine delicacies. During their immersion, sugar penetrates them to the core, then starts to recrystallise. This takes place at a particular depth and with a slightly different texture each time the temperature of the syrup is adjusted. This is what gives the marrons their sophisticated, complex consistency. Our method is positively crude compared to the traditional, sixteen-stage French technique, but it still produces delectable results.

TO MAKE MARRONS GLACÉS

English sweet chestnuts are often on the small side. You can use them to make excellent marrons glacés, but nuts from Italy, Spain, France and elsewhere are larger and very tempting.

1kg (2¼lb) large sweet chestnuts
600ml (1 pint) water (plus water for simmering the chestnuts)
600g (1¼lb) granulated sugar
200ml (7fl oz) liquid glucose/glucose syrup
2 teaspoons natural vanilla extract

Using a sharp knife, cut 2 or 3 long shallow slits in each chestnut.

Heat a pan of water to 85°C/185°F, add the sweet chestnuts and simmer them at this temperature for half an hour. You don't want to boil them as this would harden them.

Remove the pan from the heat and, as soon as the chestnuts are cool enough to handle, peel them one by one. You need to remove the outer shells but also the inner membranes. Use a paring knife and be very careful. If some of the nuts disintegrate, never mind.

If you are lucky, you will now have 30–40 pristine chestnuts. Place them in a single layer on a shallow, heat-resistant dish. Now combine the sugar and the water in a pan and boil until the temperature reaches 104°C/220°F. Pour the syrup over the chestnuts and leave in a cool place for 24 hours.

The following day, drain the syrup back into a pan and heat it to 110°C/230°F, a few degrees higher than before. Pour over the chestnuts and leave them to steep overnight again.

For the final stage, drain the syrup into a pan as before, but this time add the liquid glucose/glucose syrup and the natural vanilla extract. Heat to 116°C/240°F (softball), pour over the chestnuts and transfer them into a sterilised jar and seal (see page 164).

These marrons will keep in a cool place for an eternity. Nick uses them to make luxurious chocolate brownies.

CRYSTALLISED VIOLETS

During the Middle Ages, crystallised violets (and rose petals) were a popular tonic for sick royal children. They will be familiar to many people as the sparkly, vivid-blue nuggets that sit on top of violet creams in upmarket boxes of chocolates. In southern France you can buy them by the bagful. They are crunchy and aromatic and the colour alone is enough to perk anyone up.

The two best varieties for crystallising purposes are the sweet violet (*Viola odorata*) and the even more fragrant Parma violet. The plants are readily available in garden centres. Stay clear of yellow cultivars, in case you are tempted. Making the finished article is almost laughably simple.

TO MAKE CRYSTALLISED VIOLETS

- 20 fresh undamaged violet flowers, including a length of stalk (this is important for dipping purposes)
- 1 egg white, whipped until frothy but not stiff
- 1 tablespoon icing sugar

Pick up a violet by the stalk and dip it in the egg white, or paint on the white with a brush. Gently shake off the excess egg, then sift icing sugar over the flower while twirling it between your thumb and forefinger. The thin layer of sugar that adheres to the petals will be absorbed and form a crystalline crust.

Repeat the dipping and sugaring with all the violets, placing them on kitchen paper when done. Transfer them to the fridge, still sitting on the paper, and leave them there for a day. Then move them to a warm place like an airing cupboard and leave for a further 24 hours. The sugar should now have set.

At this stage you may want to separate the stems from the petals and discard them as their work is done.

The crystallised flowers will keep for 8 weeks in an airtight container. Use them as decorations for cakes, puddings and chocolates.

PUMPKIN AND MAPLE SPREAD

Pumpkins are tremendous items, but let's face it, they can be one of life's problems. They're enormous, won't fit in your fridge's vegetable compartment, and most families find themselves with more orange pulp than they know what to do with at least once a year – at Halloween, and if they're American, at Thanksgiving too.

Behold the solution. Struck by an inspiration, Nick managed to cadge some syrup from his father's errant brother Reuben, last seen stowing himself away on a ship to Canada some 50 years ago, but since rumoured to have become a maple magnate. The result is a magnificent amber spread, dark, rich, grainy and earthy.

1.5 kg (3lb 5oz) orange pumpkin, peeled and
roughly diced
600g (1¼lb) maple syrup
300g (10½oz) clear honey
A couple of cinnamon sticks (optional)

Immerse enough storage jars to hold slightly less than 2 litres (3½ pints) of the finished product in boiling water for 10 minutes, together with their lids. Leave them in the water until you need them.

In a thick-bottomed saucepan, gently heat all the ingredients. When the liquid starts to simmer, froth will develop on the surface. Skim this off with a ladle to keep the liquid clear. Gently boil for about 1½ hours. The pumpkin will become translucent. If any further froth appears, skim immediately. At the end of the cooking, discard the cinnamon sticks if using.

At this point, you could simply decant the product, because pumpkin preserved in this way is delicious. But we tend to blend it because the result is more versatile.

Pour into the jars immediately and seal (see page 164). The spread should keep for at least 6 months.

WAYS TO USE IT INCLUDE:
a) spread on toast with lashings of butter.
b) eaten with toasted waffles and whipped cream.
c) combined with cream and eggs to make a delicious pie or tart filling.
d) eaten neat when feeling low.

PUMPKIN, MAPLE AND RASPBERRY TART

The god of the hearth moves in strange ways. This recipe was triggered by a table-tennis accident, in which Johnny tripped over one of Nick's giant pumpkins and continued into the raspberry patch. Fearful of litigation and finding himself with a lot of soft fruit and orange pulp on his hands, Nick devised this excellent tart to mollify him. **Makes two tarts, each serving 6**

THE 'JAM'
See the instructions on page 161

THE PASTRY
500g (1lb 20z) '00' grade plain flour
175g (6oz) icing sugar
250g (9oz) butter
1 medium egg, plus 2 egg yolks
½ teaspoon natural vanilla extract

THE FILLING (TO FILL
2 TART SHELLS)
300ml (½ pint) double cream
4 medium eggs
8 yolks
600ml (1 pint) pumpkin and maple spread (see page 161)
160g (5½oz) raspberries

- To make the pastry, first sift the flour and icing sugar into a large bowl. Work in the butter with your fingers until the mixture is soft and crumbly.
- Combine the egg, egg yolks and vanilla, then make a well in the flour and pour the mixture in. Gradually stir in the flour, taking a little more into the centre with every rotation of the spoon.
- Loosely press the pastry out and divide it into two balls. Cover with clingfilm and leave in the fridge for 30 minutes before you want to use it.
- Preheat the oven to 190°C/375°F/gas mark 5. Take two 20cm (8in) flan tins with removable bases, approx 4cm (1½in) deep. Roll out the pastry and press out into the tins, so that a little hangs over the lip. Weigh down the pastry by cutting out a circle of greaseproof paper, placing it on the uncooked pastry and putting some beans or pulses on top.

- Bake in the oven for 10 minutes, then remove the beans, trim the pastry with a sharp knife and cook for a further 15 minutes or until golden.
- To make the filling, whisk the cream, eggs, egg yolks and pumpkin and maple spread together with a whisk for 30 seconds, then add the raspberries.
- The best way to get the filling into the shells is to spoon it in while the shells are still resting on the oven shelf. Fill them to within 5mm (¼in) of the top and bake for 1 hour at 150–160°C/ 300–325°F/gas mark 2–3. Check the tarts after 45 minutes. You should see the filling start to set from the sides inwards. As soon as this process is complete, the tarts are ready.

JAMS AND JELLIES: PRINCIPLES AND EQUIPMENT

Jams are made with whole fruits boiled in sugar, whereas jellies are based on their juices alone. The crucial figure in jam-making is 104°C/220°F, also known as 'setting point'. At this temperature, the sugar in which the fruit is heated develops a new consistency. To test whether a jam has reached setting point, pour a little syrup onto a pre-chilled plate, leave it to cool for a moment, then push a trail through the middle of the mixture with your finger. If the jam crinkles and the two halves remain separate, it has reached setting point. If these things don't happen, boil on for another 5 minutes, then try again.

Jellies come in many enticing colours. Their rigidity is derived from a naturally occurring setting agent called pectin, which also plays an important role in jam-making. Pectin is abundant in some fruits, particularly in their seeds, and to make jelly from these species you don't need to add any extra. With lower-pectin varieties, you usually do.

Lemon juice or vinegar are frequently added to jellies and jams. The acids they contain help extract the pectin from the fruit and they discourage the sugar from regranulating.

STERILISING JARS

To sterilise jars or bottles, wash them in soapy water, rinse thoroughly, then immerse them in boiling water for 10 minutes before drying in a cool or recently switched-off oven. Ditto lids, seals and funnels.

FILLING JAMS AND JELLIES INTO JARS

You should do this while the jars are still hot from the sterilisation process (see above). Ladle or funnel in the still-warm jam or jelly, filling each jar to within 1cm (½in) of the top. Wipe the rim and gently smooth a disc of waxed paper, cut to fit the aperture of the jar, onto the surface of the jam. Then screw on the lid, having sterilised it in boiling water beforehand.

SEALING JARS

To seal a jar that lacks a lid, take a disc of cellophane somewhat larger than the mouth of the jar, wet one side of it with a damp cloth and place it over the top of the jar. Secure it with an elastic band. As the cellophane dries, it will shrink to form a tight seal.

Label and date the jars and store them in a cool, dark larder or cupboard.

WHAT YOU WILL NEED

The following is a list of equipment you will find either essential for jam- and jelly-making, or in the last two cases, merely very useful:

- Large, thick-based stainless-steel pan, preferably a preserving pan
- Sugar thermometer
- Sterilised jars
- Lids (bare metal will react with fruit acids and so should be avoided, if necessary, cut out some card or waxed paper and insert into the lids as barriers)
- Pectin powder
- Jelly bags/muslin
- Wax paper
- Cellophane sheets
- Jam funnel
- Labels

STRAWBERRY JAM

Toast with strawberry jam has long been a pillar of British culture. This is a timeless recipe, worthy of any fête or vicar's tea-party.

Strawberries are low in pectin, but there is no need to add any here as we're not aiming for a full set. There is some pectin in the seeds and this is extracted by the lemon juice. Meanwhile, the syrup is reduced to give it a thicker consistency. But the crucial textural component of this jam is the fruit itself. There's nothing like biting into a soft, sweet strawberry through its slightly leathery skin half way through eating your toast.

Your strawberries should be as fresh as possible, preferably just picked. Above all, they should be in season. Forced or imported strawberries lack the candyfloss scent of ripe domestic ones and they make for bland jam.

> 3kg (6¾lb) strawberries, hulled
> Juice of 2 lemons
> 2.5kg (5½lb) white sugar

Incorporate your smaller berries whole and chop the larger ones into chunks. Place them in a large saucepan with the lemon juice and gently boil for an hour or so until the volume has reduced by about 10 per cent.

Add the sugar and continue to boil until the temperature rises to setting point (104°C/220°F). Carry out the setting test (see page 164).

When the jam is ready, skim any scum from the top, then remove it from the heat and leave until a skin has formed on the surface. Now stir it and pour it into jars, using a jam funnel if you have one.

Seal the jars (see page 164) and store them in a cool place. They will keep for at least 1 year.

FIG JAM

Johnny was recently faced with a painful choice when he moved into a house with a giant fig bush in the garden. Should he trim it back or give up table-tennis? He decided to get out the clippers, but because figs grow in a two-year cycle, he was forced to sacrifice most of the following year's crop. The next summer was the hottest and sunniest in history and he found himself virtually figless. He now manufactures this oh-so-simple jam to make sure it never happens again.

> 1kg (2¼lb) ripe, juicy figs
> 800g (1¾lb) sugar
> Juice of 1 lemon

Rip the stalk ends from the figs, roughly chop them and place them in a pan.

Bring them to the boil in their own juice, then add the sugar and lemon juice.

Boil until the jam has reached the magical figure of 104°C/220°F. Carry out the setting test and remove from the heat if it passes. Leave to cool until a skin has formed on the jam, then funnel it into pots and seal (page 164).

PEACH COMPOTE

The best peaches are grown on vines. Being rather small and grey, they don't fit in with modern cosmetic expectations, so supermarket buyers ignore them. They are, however, unbelievably flavoursome, as Nick found out when he pestered a farmer into surrendering some. You don't have to use vine-grown peaches to make this compote, but they should be ripe and juicy and preferably freshly picked. Peach compote is extraordinarily good with ice cream.

> 1kg (2¼lb) nice, ripe juicy peaches
> 500g (1lb 2oz) sugar
> Juice of 1 lemon
> 2 tablespoons brandy
> A few green cardamon pods or a cinnamon stick (optional)

To skin the peaches, dunk in boiling water for 1 minute, then plunge into iced water. The skin should peel off easily. Now roughly dice the flesh, discarding the stone.

Boil the peaches along with the rest of the ingredients until they have reduced by 10 per cent, then pack into sterilised jars (see page 164).

Keep the compote in the fridge as it has a relatively low sugar content. It will have a shelf-life of at least 6 months.

PEACH COMPOTE CRUMBLE

If you are feeling lonely, depressed or just plain hungry, make this crumble. It is the perfect antidote to most of the awful things in life and excellent with custard. Serves 4

> 100g (3½oz) butter
> 100g (3½oz) caster sugar
> 100g (3½oz) plain flour
> 100g (3½oz) ground almonds
> 50g (2oz) dried apricots, finely diced
> 400g (14oz) peach compote (see left)
> Lots of custard (see page 183)

- Preheat the oven to 200°C/400°F/gas mark 6.
- In a large bowl, mix the butter, sugar, flour, almonds and apricots together using your (clean) fingers until all the butter has been absorbed.
- Spoon the compote into an oven dish, then cover with the crumble mix.
- Bake for 30 minutes until brown and crisp on top. Did we mention that this is rather good with custard?

CAMILLA'S DAMSON AND CRAB APPLE JELLY

This autumnal British condiment is perfect with roast meats. Damsons (which are related to plums) and crab apples both grow wild in the UK, but in this instance they were scrounged from Nick's friend Camilla. The sugar in the recipe counteracts their tartness to produce a lovely, savoury jelly. Both fruits are high in pectin, the natural gelling agent which gives the finished product its texture.

2kg (4½lb) diced crab apples, windfalls are best
1kg (2¼lb) damsons
Juice of 3 lemons
Sugar

Place the crab apples and damsons in a saucepan and simmer for at least 2 hours. The fruit should be thoroughly pulpy.

Take a muslin cloth or jelly bag, mould it into a bowl large enough to fit the contents of the saucepan, then pour in the boiled fruit. Pick up each corner of the muslin and make a 'sack' containing the fruit pulp. Hang it from a hook over the bowl and leave overnight.

In the morning, give the muslin a little squeeze to see how much liquid is left in the cooked fruit. Text books tell you never to do this or you may get a cloudy end product, but we've found this to be not entirely true. In any case, you want to maximise the juice yield.

Measure the juice and add an equal amount of sugar by weight, in other words 1kg for every litre of juice (2¼lb for every 1¾pints) and the lemon juice. Heat the mixture to boiling point, stirring constantly, and boil for 1 minute. Skim the jelly, pour immediately into sterilised jars and seal (see page 164).

Store the jars in a cellar or larder and consume within 1 year.

REDCURRANT JELLY

In Britain, it is mandatory to accompany roast lamb with redcurrant jelly (in addition to mint sauce). 'Jam with lamb!' the French sometimes snort in derision. But they do have the concept of *sucré-salé*, or sweet-savoury, so they really should give it a try.

This recipe produces a firm, jewel-like jelly, excellent not only with lamb but also with cheese and as the basis for Cumberland sauce. Redcurrants are full of pectin-stuffed pips, so there's no need to add any to encourage setting.

For every kilo (2¼lb)of fruit, you'll be lucky if you end up with 450g (1lb) of redcurrant jelly.

> **At least 1kg (2¼lb) redcurrants**
> **sugar weighing 60 per cent of the weight of the**
>   **redcurrants (leaves a margin for error)**
> **port (optional)**

Heat the redcurrants in a saucepan and gently simmer for 1 hour. Then take a potato masher and give the fruit a good pulping in the pan.

Strain the mush through a jelly bag or muslin. Leave it to drip for several hours (see page 167).

Measure the collected juice, and for every litre add 1.2kg of sugar (1½lb to the pint). Bring to boiling point, stirring constantly to dissolve the sugar, then boil on for a few minutes until the liquid reaches setting point (104°C/220°F). At this stage, if you like, you can add a glass of port to the mixture, in which case bring the liquid briefly back to the boil.

Skim the jelly and funnel immediately into a sterilised jar, then seal (see page 164). The jelly will keep for 1 year in the larder, but eat within 1 month once you've opened the jar and keep refrigerated in the meantime.

MINT JELLY

By and large, the French see the British habit of serving lamb with vinegary mint-based sauces as insanity. They shudder at the thought of the acetic acid interfering with the taste of their wine. In the case of old-style cafe mint sauce, made with dried leaves and malt vinegar, we'd have to concede them the point. But not with this fragrant jelly. The sugar preserves and complements the bite of the fresh mint and the vinegar enhances it.

Don't expect your jelly to turn out green. It is more likely to be pinkish.

> **100ml (3½fl oz) raspberry vinegar (see page 137)**
> **100g (3½oz) sugar**
> **1 teaspoon powdered pectin**
> **100ml (3½fl oz) white wine**
> **A large handful of mint leaves**

Heat the raspberry vinegar in a pan to boiling point and reduce in volume by about 25 per cent.

Mix the sugar with the pectin, then whisk into the vinegar until dissolved.

Add the white wine and simmer for 5 minutes.

Finely chop the mint and add to the mixture. Simmer for 1 minute, then funnel the jelly into a sterilised jar and seal (see page 164). Keep in the fridge and use within 1 month.

QUINCE JELLY

The quince has fallen from favour. This is probably because it looks like a lumpy, obese, yellow pear and tastes horribly sour when raw. But it has a glorious history. When the Ancient Greeks spoke of 'golden apples' in their myths, they were almost certainly referring to quinces (the word for 'apple' was used of fruit trees generically), which they gave to new brides as fertility symbols. This association persisted for thousands of years; the quince is a frequent motif in medieval and renaissance art. Charlemagne ordered the French to grow more of them and in 1275 Edward I had quince trees planted in the Tower of London. Chaucer mentioned the fruits, using the old French *coines*, and later on they would give the world marmalade (see page 174).

The reason for this exalted status was that the quince is the preserving fruit *par excellence*. It is bursting with pectin, which makes it a natural candidate for jelly. And when quinces are cooked, they lose their astringency and release their musky flavour. These were highly desirable qualities in times when home-preserving was more necessity than luxury and people were more patient.

Johnny has a quince tree in his garden, a nice link with the past. It is of the ornamental kind, but you can still use the fruit to make decent jelly. You are best off, however, with an eating variety.

2kg (4½lb) quinces
Sugar
Juice of 2 medium lemons

Chop the quinces into small pieces. They are very hard. Nick's friend Penelope, who knows all about them, uses a small hatchet. Don't discard the pips, which are full of pectin.

Place the quinces in a large pan with about 600ml (1 pint) water. Bring to the boil, then gently cook out, stirring frequently, until the consistency is like thin apple purée. The quinces may look as if they're never going to break down, but they will.

Strain the mixture overnight through a jelly bag (see page 167). In the morning, 'squoosh' the bag to extract maximum juice.

Measure the juice and return it to the pan. Add an equal quantity of sugar (1kg per litre) plus the lemon juice. Bring to the boil, stirring constantly. As the sugar dissolves, the juice will clarify from a greeny-grey sludge into a beautiful rosy pink. After 5 minutes, remove the pan from the heat and skim the liquid. Leave it to stand for 5 minutes, then skim it again. Then funnel into sterilised pots and seal (see page 164).

As the jelly ages it will become translucent amber. It will keep for 1 year in a cupboard or larder or for at least 1 month in the fridge after opening.

MANGO CHUTNEY

Mango chutney is so ubiquitous in the UK that people fail to appreciate how exotic it is. *Chatni*, which translates approximately to 'finger licking', was originally an Indian relish made from fresh fruits and spices. British colonists picked up a taste for these 'chutneys', brought them back to Europe and adapted the recipes. Then they spread the habit throughout the Empire.

Good chutneys can be built around peaches, bananas, apricots, plums, cucumbers, apples, damsons or tomatoes, but the mother of them all has to be the mango. It is almost unthinkable to eat curry without it, and cheeseboards always benefit from its presence.

TO MAKE CHUTNEY

Traditional mango chutney is made with salted, pickled green mangoes. These are not easy to obtain, however, so we use firm yellow mangoes instead. They make for a sweeter, better-coloured chutney.

> 1kg (2¼lb) mangoes, diced
> 50g (2oz) fresh root ginger, cut into a fine julienne
> 50g (2oz) medium-strength red chilli, seeds
>   removed, cut into a fine julienne
> 250ml (9fl oz) cider vinegar
> 200g (7oz) golden caster sugar

Place all the ingredients in a pan and bring to the boil. Simmer for 1 hour, then pack into sterilised containers and seal (see page 164) and store in a cool place for up to 1 year.

QUINCE CHEESE

This is no more a cheese than fruit leather is leather, but it does have a texture similar to semi-soft cheese. In fact, its scented, honey-like flavour goes very well with it. It is also delicious with cold meats or foie gras. As it ages, quince cheese becomes firmer, darker and more intense in flavour.

2kg (4½lb) quinces
Juice of 1 lemon
Sugar, equal in weight to the quince flesh after coring and peeling

Core and peel the quinces, reserving the flesh. Heat the cores and peelings in a saucepan with 250ml (9fl oz) water and simmer for an hour. Pass the pulp through a conical sieve and reserve.

Weigh the quince flesh, then place it in a thick-bottomed saucepan with the sieved pulp, lemon juice and another 250ml (9fl oz) water and simmer for 1 hour.

Blend the pulp with a hand-blender or food-processor until smooth.

Now add the sugar. Slowly cook the purée for another 2 hours, stirring occasionally, until it has darkened to a deep rose. The cheese is ready when it is so thick that a spoon drawn across the base of the pan leaves a definite path in its wake.

Lightly grease a shallow, non-stick baking tray and pour the cheese in to a depth of about 3cm (1¼in). Leave it to set overnight, then turn it out onto a length of muslin and wrap the material round it. Then fold a couple of sheets of newspaper over the whole kaboodle and store in a dark place for 1 month before eating. The quince cheese will keep for up to 1 year.

LEMON CURD

A good lemon curd is fresh tasting and packed with the essence of the fruit. The richness of the ingredients means that it only has a limited shelf-life, but we're still talking 6 weeks in the fridge if properly sealed.

Home-made lemon curd is invariably nicer than the shop-bought equivalent. It is excellent on toast and of course in lemon meringue pies. For many, eating curd made according to this recipe is like stumbling on a long-forgotten treasure.

4 large or 5 medium lemons, zested and juiced (use unwaxed lemons and large juicy ones at that)
4 large eggs
370g (13oz) caster sugar
250g (9oz) unsalted butter, cut into small cubes
1 level tablespoon corn flour
3x 300ml (½pint) jars
3 wax discs to fit the mouth of the jars (available from jam and catering shops)

Sterilise the jars (see page 164) and keep them to hand along with the wax discs.

Pour all the ingredients into a saucepan and whisk vigorously for 30 seconds.

Heat over a low flame, whisking constantly to make sure the contents don't stick to the side of the pan. A food thermometer is useful here. The mixture hitting 70°C/158°F is a sign that the eggs and corn flour are about to emulsify and gelatinise.

Once the mixture has thickened, carry on cooking for just another minute, then remove the pan from the heat.

Pour the curd into the jars with a funnel. Lay the wax discs over the tops, then seal (see page 164). Leave to cool, then store in the fridge for up to 6 weeks.

LEMON MERINGUE PIE

A good lemon meringue pie is a lovesome thing, but as life sadly teaches us, there is plenty of scope for messing it up. Institutions of all kinds, from schools to office canteens, are masters at ruining them. Bland, over-sweet or suspiciously coloured curd is adorned with unset, soapy meringue. This may be why the pleasure is so deep when you eat an authentic, home-made LMP. **Makes 2 tarts, each serving 4–6**

THE PASTRY

500g (1lb 2oz) '00' grade plain flour

175g (6oz) icing sugar

Zest of 1 lemon

250g (9oz) butter

1 medium egg, plus 2 egg yolks

½ teaspoon natural vanilla extract

THE MERINGUE AND FINAL TOUCHES

4 medium egg whites, plus 3 egg yolks

100g (3½oz) caster sugar

50ml (2fl oz) double cream

450ml (16fl oz) lemon curd (see page 171)

75g (3oz) icing sugar

- Make the curd according to the instructions on page 171, or grab a jar from the fridge.
- To make the pastry, first sift the flour and icing sugar into a large bowl. Add the lemon zest, then work in the butter with your fingers until soft and crumbly.
- Mix the eggs and egg yolks and vanilla together, make a well in the flour, then stir in the egg mixture.
- Press the pastry out loosely, then divide into two balls. Cover in clingfilm and leave to rest in the fridge for 30 minutes before use.
- Preheat the oven to 190°C/375°F/gas mark 5.
- Take two 20cm (8in) flan tins with removable bases, 2.5cm (1in) or 4cm (1½in) in depth. Roll out the pastry and press it into the tins, so that it hangs slightly over the lip. Weigh it down by cutting out circles of greaseproof paper, placing them on the uncooked pastry and evenly placing dried beans or pulses on top.
- Bake in the oven for 10 minutes, then remove the beans, trim the pastry with a sharp knife and cook for a further 15 minutes or until golden.

- To make the meringue topping, simply whisk the egg whites and caster sugar until the mixture forms stiff peaks. This is easily done in a mixer. If you are doing it by hand, first whisk the egg whites until stiff, then add and whisk in the caster sugar a little at a time. The mixture should become glossy.
- Now you must put the whole thing together. Stir the egg yolks and double cream into the lemon curd. Sift in the icing sugar and whisk it in. Then pour the whole mixture into the tart shells. Then cover the filling with the egg white and sugar foam. The easiest and neatest way to do this is through a piping bag, but if you don't have the equipment, just spoon on the meringue topping and smooth it with a palette knife.
- Bake in the oven at 160°C/325°F/gas mark 3 for 40 minutes. The pies are ready when the meringue is crispy and slightly coloured on top and soft in the middle.

NICK'S MARMALADE

To those of us accustomed to the orange kind, or lime if we're feeling radical, it seems strange that marmalade was originally made from quinces. It takes its name from *marmelo*, the Portuguese word for that fruit. Significant quantities of 'marmelada' were imported into Britain during the fifteenth century, but it was a couple of centuries later, when people started to make it with bitter Seville oranges, that its popularity really soared.

The USP of our marmalade is the fineness of its solid component. Nick is no big fan of big shards of peel in his marmalade, preferring tiny particles of zest lost in a golden expanse of jelly.

For this recipe you will need a very large saucepan with a capacity of around 10 litres (2¼ gallons). This is a lot of marmalade, but you'll find yourself giving most of it away to pleading friends and relatives.

TO FILL QUITE A FEW JARS:
  5kg (11lb) Seville oranges
  7 lemons
  2 litres (3½ pints) orange juice (in addition to the
    juice you squeeze out of the oranges)
  Water to top up
  Lots of sugar
  Lots of muslin

Zest the oranges with a zester or the finest face of a cheese-grater. This can be time consuming, but no more than the main alternative, which is to cut the peel into shreds. Spread the procedure over a couple of days if you like. The zest will keep in the fridge for up to 4 days.

Squeeze the lemons and scoop out their innards, pith, pips and all. Discard the peel and reserve the rest.

Cut the oranges in half and squeeze the juice. Put all parts of them except for the reserved zest into the saucepan, adding the lemon juice, pith and pips. Press down the orange halves, then pour in the supplementary orange juice. Top up with water until the liquid just seeps over the top of the oranges.

Bring to the boil, then simmer for 1 hour with the lid on.

Remove the saucepan from the heat and leave it to cool. After 24 hours, return the marmalade to the heat and gently boil for 2 hours.

Transfer the contents of the pan into the centre of an extremely large square of muslin. Squeeze and manipulate it to extract as much juice as possible.

Measure the collected juice and return it to the pan, adding 900g (2lb) sugar for every litre (1¾ pints). Bring the juice back to the boil and now add the reserved zest. Keep boiling until the mixture has reached setting point (104°C/220°F). Carry out the setting test to reassure yourself that it will set (see page 164).

Funnel the marmalade into sterilised jars, then seal them (see page 164) and label them.

Store in a larder or cupboard for up to 1 year.

ORIENTAL PLUM SAUCE

This is a home-made version of the sweet, treacly sauce served with crispy duck and pancakes in Chinese restaurants.

 4 tablespoons sunflower oil
 2 large onions, finely chopped
 2 large knobs of ginger, finely chopped
 3 cloves of garlic, chopped
 1 medium-strength red chilli, chopped
 2.5kg (5½lb) dark plums, stoned
 1 litre (1¾ pints) red wine vinegar
 300ml (½ pint) water
 250g (9oz) soft brown sugar
 200ml (7fl oz) dark soy sauce
 ½ teaspoon ground cinnamon
 1 spice bag (4 star anise and 1 teaspoon Sichuan pepper, all crushed and tied up in muslin)
 Salt to taste

Take a large, thick-bottomed saucepan and pour in the oil. Cook the onions, ginger, garlic and chilli until nice and soft, then add the plums, vinegar and water. Bring slowly to the boil, then simmer until the plums are also soft.

Push the ingredients through a sieve using a wooden spoon, then return the purée to the pan. Add the sugar, soy sauce, cinnamon and the spice bag.

Bring the sauce back to the boil. For the next 2 hours you need to stir it constantly, breaking only to sterilise half a dozen jam jars.

When the sauce is ready, take out the spice bag and season with salt to taste. Immediately fill the jars almost to the top. Seal tightly (see page 164), then leave to settle in a cool place.

The sauce/jam will keep for several months, but after opening a jar you must store it in the fridge.

DUCK WITH PLUM SAUCE

This is the classic way of using plum sauce, rolled up in pancakes with shards of roast duck and strips of cucumber and spring onion. Serves 4

5 whole star anise
5 cloves
4 cloves of garlic, peeled
1 duck, about 2kg (4½lb)
1 tablespoon brown sugar
100ml (3½fl oz) dark soy sauce

Chinese pancakes (available in Asian supermarkets, allow 3 or 4 per head)
plum sauce (see left)
½ cucumber, cut into strips
6 spring onions, cut into strips

- Insert the star anise, cloves and garlic into the duck's cavity, then roughly sew it up. Hold the bird over the sink and pour boiling water over it to tighten the skin and remove fats from the surface. Then hang it in an airy place and leave to dry for 12 hours.
- Mix the brown sugar into the soy sauce and paint the mixture onto the duck. Leave to dry for a couple of hours, then repeat the process.
- Preheat the oven to 200°C/400°F/gas mark 6 and place the duck directly onto the oven shelf. Put a dish filled with water on the base of the oven to collect the grease. Roast the duck for 1½ hours.
- Remove the meat from the bones and shred it a bit with a fork.
- Heat the pancakes in a steamer or microwave. Spread them with a little plum sauce. Place a few spring onions and cucumber strips in the middle of each pancake, then top off with a sprinkling of duck and crispy skin. Roll into a tube and take a large bite. Try to eat more than anyone else.

SWEET CHILLI SAUCE

This finger-licking sauce is a crucial component of the smoked salmon and noodle recipe on page 65, but it is also just the job with spare ribs (page 151).

> 50g (2oz) fresh red chillies, de-seeded and thinly
>   sliced
> Zest and juice of 2 limes
> 50ml (2fl oz) rice vinegar
> 40g (1½oz) fish sauce
> Splash of soy sauce
> 60ml (2½fl oz) water
> 1 level teaspoon powdered pectin
> 75g (3oz) sugar

Place all the ingredients except the pectin and 1 tablespoon of the sugar in a saucepan and simmer gently for 25 minutes with the lid off.

Mix the pectin with the remaining sugar and sprinkle into the sauce while whisking.

Simmer for a further 5 minutes, stirring gently.

Decant into sterilised pots and seal (see page 164) and allow to cool at room temperature. Then refrigerate.

The sauce will keep for 6 months in the fridge, but consume within 3 months of opening a pot.

MINCEMEAT

As a child, Johnny remembers being hopelessly confused about the relationship between mincemeat, mince, meat and mints. It was only when he learned that mince pies really had once been made with minced meat that he began to relax. You may be relieved to know that we aren't about to revive the practice, although the cow does continue to play a role by providing the suet.

TO FILL 5-6 JARS
> 4 medium Bramley apples
> Zest and juice of 3 lemons
> 500g (1lb 2oz) raisins
> 500g (1lb 2oz) currants
> 500g (1lb 2oz) suet
> 1kg (2¼lb) soft brown sugar
> 100g (3½oz) candied orange peel (see page 156),
>   finely chopped
> 50g (2oz) almonds, chopped
> ¼ teaspoon ground cinnamon
> ¼ teaspoon ground ginger
> 125ml (4fl oz) brandy
> 100ml (3½fl oz) Grand Marnier or other orange liqueur

Core the apples and place them in a lidded dish. Bake for 1 hour or until soft in an oven heated to 200°C/400°F/gas mark 6. Squeeze all the pulp from the apple skins into a large bowl. Add the lemon juice and zest, then stir in the rest of the ingredients.

Cover the mince and leave it in a cool place for a couple of days. Give it an occasional stir.

Fill the mincemeat into sterilised pots or jars (see page 164), taking care to remove any air bubbles. Seal (see page 164) and store in the larder. It will improve with age. Wait 2 months before eating it.

ALCOHOL

Many of the world's favourite alcoholic drinks are made with fruits, so it seems particularly appropriate to use it to preserve them. Alcohol is itself a product of a preserving process (fermentation) and it in turn has a powerful embalming effect on food immersed in it. Nothing can grow in pure alcohol, which is why it is used by entomologists and botanists to keep biological specimens intact. For this reason, it is usually best to use the strongest available brand of the relevant liquor for preserving purposes.

There is more truth than you might think in the old story of the priest who decided to preach a sermon on the evils of booze. First he put an earthworm in a glass of water, where it wriggled about quite happily. Then he placed it in a tumbler of whisky. It dropped to the bottom stone dead. 'And what is the lesson here?' he asked his congregation. 'I know,' shouted out an old drunk who had blundered into the church. 'Water gives you worms.'

In the Middle Ages, when water supplies were untreated, drinking 'Adam's ale' could pose a serious health hazard. In this context, brewing and distilling were not simply luxuries; they were also forms of hygiene. Even children were given 'small beer'. Monks were particularly expert at making alcohol and flavouring it with fruits and herbs. They also pioneered the practice of steeping the fruits they grew in their orchards in spirits and liqueurs to prolong their shelf-lives. The end-products were warming and cheering, especially in winter. They also occupied an important niche in the humoural system of medicine.

Fruits preserved in alcohol became so popular during the eighteenth century that many varieties, including peaches, apricots, cherries and nectarines, went by the generic name of 'brandy fruits'. They were typically served in dedicated glasses at the end of indulgent Georgian meals.

When fruits are immersed in alcohol, they absorb its flavours and vice versa. The sugars they contain are also encouraged to ferment, particularly if extra sugar is added. This is the principle behind the manufacture of Sloe Gin (see opposite). Aside from the kick imparted, preserving fruits in alcohol has many advantages over other methods. There is no need to use additives, no loss of vitamins through heat treatment and the fruit stays firm. The results are also beautiful and go very well with ice-cream.

SLOE GIN

Sloes, the fruit of the spiky blackthorn bush, grow wild along many paths and hedgerows in the UK. They ripen in the autumn and are like hard, miniature, vividly blue plums, to which they are closely related. Sloes impart a wonderful, warming fruitiness and colour to gin or vodka. You wouldn't want to eat them raw as they are far too bitter, but numerous birds think otherwise. You will be in direct competition with them for the harvest.

Tradition has it that the best time to pick sloes is after the first frost of autumn has swollen and softened them slightly. You might want to wear gloves as blackthorns are notoriously prickly.

TO MAKE ENOUGH TO SEE YOU THROUGH TO THE FOLLOWING AUTUMN
  2kg (4½lb) sloes (a good afternoon's harvest)
  1kg (2¼lb) sugar
  3 bottles of gin or vodka, the higher proof, the better

The onerous part is pricking each fruit several times with a needle, or, more traditionally, a blackthorn spike. Either way, settle down for a couple of meditative hours. Place the pricked sloes in a large sealable jar, pour in the sugar, then add the alcohol. Seal the jar and invert several times to distribute the sugar and start it dissolving.

Leave the gin/vodka to mature in a cool, dark place for 6 months. Turn it over occasionally, particularly in the early stages.

After this period, strain the liquid through muslin or clean cotton cloth and decant it into sterilised bottles (see page 164). If you're canny, you will have kept the original spirit bottles and their caps. Wait 6 months to 1 year before drinking.

MORELLO CHERRIES IN KIRSCH

The dark morello cherries that grow in abundance in the Black Forest in Germany are too sour to simply eat from the tree. Instead, the locals mash them in large wooden tubs and allow them to ferment. The end result is *kirsch* or *kirschwasser*, a dry, colourless hooch with quite a kick to it. Fresh cherries steeped in the kirsch become eminently edible. Their sourness is replaced by a luxurious tangy sweetness, and they are immediately co-opted into a *Schwarzwalderkirchtorte* (Black Forest cake).

1kg (2¼lb) morello cherries
350g (12oz) sugar
300ml (11fl oz) water
150ml (¼ pint) brandy
150ml (¼ pint) kirsch

Cut slits down the sides of the cherries with a paring knife, then ease out the stones. Alternatively, leave the stones in but cut a cross in each cherry at the stalk end to let the juices out and the alcohol in.

Add 100g (3½oz) of the sugar to the water in a pan and bring to the boil. Add the cherries and simmer for 5 minutes.

Remove the cherries with a slotted spoon and reserve. Add the rest of the sugar to the syrup, dissolve and boil gently for 5 minutes. Leave to cool.

Drop the cherries into a sterilised preserving jar (see page 164) until filled to the neck. Then combine the brandy and kirsch with the syrup and pour it over the fruit. Seal the jar (see page 164) and wait at least 3 months before use.

These cherries are the defining ingredient of the morello cherry trifle opposite.

TRIFLE WITH MORELLO CHERRIES

There isn't enough space here to fully extol the virtues of this trifle, but consider it the last word on the subject. Serves 6

THE CUSTARD

100g (3½oz) caster sugar
650ml (23fl oz) full-fat milk
100ml (3½fl oz) double
  cream
20g (¾oz) corn flour
1 teaspoon natural vanilla
  extract
Pinch of vanilla seeds
125g (4½oz) egg yolk,
  lightly beaten

THE SPONGE

150g (5oz) unsalted butter,
  softened
150g (6oz) caster sugar
2 large eggs, plus
  4 large egg yolks
Few drops of natural vanilla
  extract
250g (9oz) '00' grade
  (extra fine) flour
2½ teaspoons baking
  powder
Large pinch of salt

100ml (3½fl oz) milk
100ml (3½fl oz) cream
23cm (9in) cake tin with
  removable base

THE SYRUP AND THE CHERRIES

300g (10½oz) jar morello
  cherries preserved in
  kirsch (see left)

THE TRIMMINGS

75g (3oz) shelled
  pistachios, lightly roasted
  in the oven for 5 minutes
  at 190°C/375°F/
  gas mark 5, then crushed
high-quality dark chocolate,
  for grating
400ml (14fl oz) double
  cream with 2 tablespoons
  caster sugar whipped in

6 glasses holding around
  200 ml (7fl oz) each

- To make the custard, combine the sugar, milk, cream, corn flour, vanilla extract and vanilla seeds in a thick-bottomed saucepan. Simmer for 5 minutes, whisking continuously.
- Remove from the heat and whisk in the egg yolks.
- Return to a very low heat for a minute or two, then the custard will be ready. Store it in the fridge until you need it.
- Butter the base and sides of the cake tin, then sift flour over the whole buttered area to create a non-stick barrier.
- Preheat the oven to 180–190°C/350–375°F/gas mark 4–5.
- Cream the butter, preferably with an electric mixer, until lightened in colour. Then add the sugar and continue to beat for 4–5 minutes. You may need to switch the mixer off from time to time and turn the mixture in with a spatula.
- Slowly add the eggs. If you do this too fast, the mixture will curdle. Then add the vanilla extract.
- Combine the flour with the baking powder and salt. Add a little of this mixture to the butter, sugar and eggs, then add a little of the milk and cream. Keep alternating wet and dry ingredients until they have all been incorporated.
- Stir the mixture until smooth. Then pour it into the cake tin and bake for 30–40 minutes. If the surface of the sponge starts to brown during cooking, turn the heat down a little. To check whether the sponge is ready, insert a cocktail stick into the centre. If it comes out clean, it's done.
- Push the sponge out of the tin and onto a cake rack to cool. You will have too much for this recipe, so freeze the rest.
- Pour the kirsch syrup into a pan, reserving the cherries and, reduce by half over moderate heat. Reserve and chill, along with the roughly chopped cherries.
- Slice the sponge into layers about 2cm (1in) thick, then cut it to fit your glasses. Pour over the reduced kirsch syrup, then fill each glass in the following order: cherries, custard, pistachios, sponge, then the same again, then a dollop of cream. Shave or grate a little chocolate on the summit.

ORANGES IN BRANDY

Brandy-soaked oranges were a big hit at bibulous Georgian dinner parties. Oranges were still something of rarity, which gave them status, and given their limited lifespan it made good sense to preserve them. Guests were not disappointed when they found that the juice in each segment had swapped place with the fragrant alcohol. The beauty of this dish lies in the trick it plays on the brain. The orange looks freshly cut but tastes wonderfully boozy. Each mouthful comes as a surprise to the tastebuds, even when you've consciously got the hang of the situation.

These oranges make beautiful gifts, packed tightly in their deep amber cosmoses. To vary both visual effect and flavour, try experimenting with blood oranges, ortaniques, mandarins or whatever takes your fancy.

HOW TO MAKE ORANGES ALCOHOLIC

12 small oranges
350g (12oz) sugar
300ml (½ pint) water
300ml (11fl oz) brandy
2 preserving jars around 1 litre (1¾ pints) in capacity

Remove the zest from the oranges with a grater. Simmer the zest in water for 30 seconds, then drain and reserve.

Peel the oranges, cutting into the flesh to make the globes slightly smaller than they otherwise would be. The individual sacs of juice inside the segments should be exposed.

Heat the sugar and water in a saucepan, making sure that all the sugar dissolves. Simmer the oranges in the syrup for 2 minutes, then remove. Boil the syrup for 5 more minutes, then remove from the heat and allow to cool.

Pack the oranges up to the necks of the sterilised jars (see page 164), adding a few sprinkles of the reserved zest.

Add the brandy to the syrup, then pour the liquid over the oranges, making sure they are completely covered.

Seal the jars (see page 164) and leave to mature in a cool dark place for a minimum of 2 months. Once opened, they will keep for 6 months as long as the oranges are covered by liquid.

A BOOZY ORANGE DESSERT

Nick has devised a fine dessert based on oranges in brandy. To make enough for 4 people, cut 4 of the preserved oranges into segments and reduce their preserving liquid until quite viscous (approximately a quarter of its initial volume). Meanwhile, cook 8 fresh dark plums in 50ml (2fl oz port) and 50g (2oz) sugar. Pass them through a sieve, reserving the juice. Arrange all the ingredients in flute glasses interspersed with scoops of vanilla ice-cream. Pour over the liquids and garnish with a little zest.

LAI'S FRUITS-OF-THE-FOREST-FLAVOURED RUM

There are several long, thin bottles under Nick's stairs, laid down horizontally and covered in foil. They were put there seven years ago by his now-wife Lai when she was marking out her territory. Inside them lies the nectar of gods. Occasionally, if they have had a baby or won the lottery or similar, they decant a nip or two and luxuriate in its beneficial properties. This is a welcome release from their frugal lifestyle.

Nick has been nagging Lai for the recipe for seven years. Only her sense of duty to this book made her relent. She sighed deeply and revealed her secret, 'I went to Marks & Spencer, bought loads of frozen "fruits-of-the-forest" mix, stuffed it into rum bottles and then topped it up with rum and sugar.' All the mystery has now gone out of their marriage.

Bear in mind that it is the liquid we are interested in here, not the fruit mush, which you are under no compulsion to eat.

1kg (2¼lb) frozen fruits-of-the-forest mix*
300g (11oz) sugar
1 litre (1¾ pints) dark rum
Aluminium foil to wrap the bottles
3 used 750ml spirit or liqueur bottles with their original corks or caps

*You can buy this from M&S and other supermarkets, or you can collect your own fruits and freeze them, taking a look at chapter 12 *en route*. Good fruits to use include black- and redcurrants, blueberries, blackberries, strawberries and raspberries.

Patiently insert the fruits-of-the-forest into the sterilised bottles (see page 164).

Mix the sugar with the rum and decant into the bottles via a sterilised funnel.

Cover the bottles with foil and leave for at least 1 year. It doesn't hurt to give them a little shake from time to time.

Serve neat, or try in a cocktail with champagne and fresh blackberries.

BOTTLING AND CANNING

If one invention has been more responsible than any other for the spectacular increase in the human population over the last few centuries, it is that of the microscope. Prior to this breakthrough, the true cause of the majority of diseases was anyone's guess. And this, of course, applied to food poisoning.

The real enemy when it comes to bottling and canning is the ubiquitous bacterium *Clostridium botulinum*. Its spores are everywhere. Completely avoiding them is impossible, but fortunately eating them in regular quantities is not harmful. It is only when they grow in astronomical numbers, as when they find themselves in an ideal environment like an improperly canned jar of food, that trouble ensues. As they begin to die off, they produce a neurotoxin, and even a small bite of a contaminated food can be lethal. This observation shouldn't put you off, but it should convince you to take care. Many kinds of contamination are betrayed by tell-tale signs. These include popped-up or bulging lids, surface mould and cloudy or discoloured brine. Unfortunately, this is not the case with the *Botulinum* bacteria. You can only be confident that you have killed them off or rendered them harmless if you have scrupulously adhered to the rules given here. Anyone who hasn't should seek urgent medical advice if they experience unexpected breathing difficulties after eating home-canned food.

Strangely, the immediate catalyst for the discovery of a safe way of bottling and canning food (we use the terms synonymously and aren't about to teach you welding or aluminium processing) was Napoleon. He offered a prize to the first person to come up with a method of producing long-lived, non-dried comestibles for his armies. The challenge was taken up by Nicholas Appert, who demonstrated that foods could be preserved by placing them in sterilised glass containers, then sealing and heating them. Shortly afterwards, his rival Peter Durand obtained a British patent for canning with tin-coated steel cans. Unfortunately, the solder used at the time became the source of a new problem: lead poisoning. This was what finished off Sir John Franklin's crew when, in the middle of the nineteenth century, they found themselves stranded on ice during an attempt to locate the North-West Passage. Luckily, this problem was ironed out by the invention of the double-folded canning seal.

Canned food has a dubious reputation among nutritionists as a fair amount of vitamins A, C, thiamine and riboflavin are destroyed during heating. But in fact these sensitive substances degrade very quickly anyway. If you can victuals in good condition shortly after they are harvested, they are likely to be more nutritious several months later than equivalents which have been sitting in the fridge or on a supermarket shelf for a couple of days. Leaving all this aside, the alchemy involved in home canning is enjoyable, and the results often have unique flavours and textures. We defy anyone not to like home-tinned peaches.

PRINCIPLES AND EQUIPMENT

Canning food safely requires a certain amount of specialised equipment and a great deal of care and precision, but this only enhances the satisfaction and sense of achievement when you consume the end-products months down the line.

To reiterate, the important thing when it comes to bottling or canning food is not to give everyone botulism. Most bacteria, moulds and yeasts can be killed off by boiling their host foods and the jars that contain them at the 'regular' temperature of 100°C/212°F. *Clostridium botulinum*, however, can survive this treatment, so special precautions need to be taken.

The steps you need to take to prevent contamination depend on the acidity of the items you are processing. High-acid foods, which include most fruits, can be processed under the less restrictive conditions of 'boiling-water canning' This is because they have a low-enough pH to prevent the growth of the *Botulinum* bacteria. Low-acid foods, on the other hand, do not. They include meat, fish, milk products and vegetables. To render them safe, they need either to be cooked under artificial pressure, which raises the boiling point of water, or to be boiled at normal atmospheric pressure for a ridiculously long time. If you are planning to can low-acid foods, it is much more sensible to purchase a pressure canner, which is essentially a bespoke kind of pressure cooker. But don't use the ordinary kind as they lack gauges.

JARS AND CONTAINERS

As far as containers go, a threaded glass canning jar with a self-sealing lid is the best option. This lid consists of a flat metal plate held in place by a metal screw-band. The plate is edged with a special compound which melts during processing but still allows air to escape. Then it forms an airtight seal as the jar cools. This is reinforced by the vacuum created beneath during heating.

You need to clean your jars thoroughly before they are used, but unless you are using them to store something subjected to heat for less than 10 minutes, there is no need to sterilise them in advance, as this will happen during the boiling phase.

PACKING AND SEALING JARS

Canned food can be either hot-packed, where it is simmered for a few minutes prior to packing (or, in the case of precooked items like baked beans, transferred directly from the pan or oven into the jar), or raw-packed, where it is placed in the jar without preheating. We recommend hot-packing for all the items below. Hot-packing minimises the amount of air in the food, which would otherwise cause fruit to float and other foods to discolour and deteriorate. Raw-packing is mainly suitable for pressure-canned vegetables. In either case, the liquid poured over the food should be boiling or freshly boiled.

When packing your jars, you should leave a little headspace between the surface of the food and the lid. This will cater for expansion during processing and allow a vacuum to form when the jar cools later on. The desirable headspace varies from 1.25cm (½in) for fruits to up to 2.5cm (1in) for low-acid foods to be canned under pressure.

Whichever canning method you are using, you need to remove as much air as possible from your food and its surrounding liquid before you seal the jars and subject them to heat. Before you finally put on the lid, insert a flat plastic spatula down the side of the jar and rotate the jar slowly while moving the spatula up and down to dislodge any air bubbles between the food and the walls.

If you are using preserving jars where the lids are sealed with screw-bands, wipe the rim of each after filling and cover with a sterilised lid. Then, holding the jar steady with your 'free' hand, screw down the band until tight, then give it a quarter turn the other way. Each lid should only be used once.

We'd recommend using screw-band jars for all the products in this chapter, but you can also use clamp-top/kilner jars with rubber sealing rings. Use a new ring on each occasion. Some items (e.g. tomato ketchup, page 102) are most at home in corked bottles. These should have ridges at the top to allow you to tie down the corks – this is important if they are to stay in place during heating. Fill the bottles to within 1cm (½in) of the top using a funnel and ladle. Soak the corks in hot water for a few minutes. Push them into the bottles as far as they will go, then tap them down with a mallet. Then take 50cm (20in) lengths of string and tie the corks down securely. This will be easier if you cut a shallow groove into the top of each cork.

BOILING-WATER CANNING

For boiling-water canning, you need a pan tall enough to hold enough water to more than cover your jars by an inch or two, plus an airspace of another couple of inches between the surface of the water and the lid of the canner. This will allow a vigorous boil. There is a lot to be said for purchasing a dedicated boiling-water canner. This will include a rack to prevent your jars from toppling over during boiling.

The boiling point of water falls with altitude because there is less air above pushing down on it. As a result, boiling-water canning (and for that matter boiling an egg) takes longer up a mountain than at sea level. We show you how to adjust for altitude in the instructions for the relevant items.

PRESSURE CANNING

Pressure canners allow you to heat food to the temperature required to kill botulinum spores (116–121°C/240–250°F). They vary somewhat in design according to manufacturer (take a look at the splendid 1940s model shown on page 186) but essentially consist of a large kettle with a clamped or turn-on lid featuring a steam vent or petcock. This is closed during processing to allow the pressure inside the kettle to build up. There will also be a pressure gauge. Depending on the model, this may take the form of a dial, or it may be a weighted gauge which is itself used to close the steam vent. In the latter case, the gauge will be inscribed with numbers corresponding to specific pressures. When it is placed over the steam vent, the desired pressure reading should be lined up with the mark or arrow on the lid (the manufacturer's instructions should make all this clearer). A good canner will also include a removable jar rack and a safety fuse on the lid that will blow if the internal pressure grows excessive.

Those who know their weights and measures will notice that the recommended pressure settings for processing the items below are less than normal atmospheric pressure (14.7 psi). This is because the reading on the gauge or dial refers to the additional pressure inside the canner. When it registers 1 bar/14.5 psi at sea-level, the internal pressure is therefore almost 2 atmospheres. What really matters, however, is the total pressure inside the canner. This is influenced by altitude: the higher you are, the lower the atmospheric pressure. To generate the same total pressure, you therefore need to increase the internal pressure of the canner.

Canners with weighted gauges are usually less adjustable than those with dials. You will probably only have 3 settings to choose from. If you have such a canner, set it to 0.69 bar/10 psi for all processes carried out from 0–305m (0–1000ft) and 1.035 bar/15 psi above 305m (1000ft).

Instructions for adjusting pressure-settings and boiling times at altitude are given for each item in this chapter, although we've assumed you aren't about to start canning in the Himalayas.

THE PROCEDURE FOR PRESSURE CANNING

Pour 5–7.5cm (2–3in) of hot water into the kettle, place the sealed jars in the rack and securely fasten the canner lid.

Next, heat the kettle at its highest setting with the steam vent or petcock left open.

After a few minutes, steam will start to pour out, and you should allow this to continue for 10 minutes. Only now do you close the petcock or place the weighted gauge/counterweight over the steam vent. If your gauge is the weighted kind, position it so that the desired pressure setting is lined up with the reading mark on the lid.

The canner should take about 3–5 minutes to pressurise. In the case of a weighted-gauge type, you will know that it has reached the right level when the weight itself starts to jiggle. Be not afraid, this is how it is designed to behave. Every time it jiggles, a little steam is released, regulating the internal pressure. If, on the other hand, you have a dial gauge, the kettle is pressurised when the dial says so.

Time your canning from the moment the gauge first rocks or registers the required pressure reading. Regulate the heat under the kettle to maintain steady pressure during the process.

When the jars have been processed for the length of time appropriate to their contents, turn off the heat and leave the canner to cool. Leave the petcock or steam vent well alone at this stage. The internal pressure will gradually drop as the kettle cools towards room temperature.

After 30–45 minutes, open the steam vent or petcock, wait 2 minutes, then unfasten the lid and remove it, taking care not to be hit by the blast of steam.

Remove the jars, leave them to cool and store.

COOLING

Allow jars/bottles to cool at room temperature for 12–24 hours.

ILLUSTRATIVE EXAMPLE – TOMATO PASSATA

You can use either method to can tomato passata (see page 195), so it serves as a useful example of the many variables to be considered when calculating processing times and pressure settings. Depending on your altitude, the size of your jars and the canning method you are using, these should be as follows:

BOILING-WATER CANNING

½ litre/pint jars: 35 minutes at 0–305m (0–1000 ft)
    40 minutes at 305–915m (1000-3000ft)
    45 minutes at 915–1525m (3000–5000ft)
and so on.
Litre/quart jars: add 10 minutes to the processing time.

PRESSURE CANNING

½ litre/pint jars: 20 minutes at
    0.42 bar/6 psi from sea level–710m (0-2000ft)
    0.48 bar/7 psi from 710–1420m (2000–4000ft)
    0.55 bar/8 psi from 1420–2130m (4000–6000ft)
and so on.

Litre/quart jars: 15 minutes at:
    0.76 bar/11 psi from sea level–710m (0–2000ft)
    0.83 bar/12 psi at 710–1420m (2000–4000ft)
    0.9 bar/13 psi from 1420–2130m (4000–6000ft)
and so on.

These figures are specific to tomato passata. The recommended processing times and pressure settings for the canning of the other items in this chapter are given on a case-by-case basis.

SAFETY AND STORAGE

Canned foods are best kept in dry, dark places that are relatively but not excessively cool. These conditions help to preserve vitamins and taste and minimise the chances of spoilage.

To check the seal on a processed jar, gently undo the clamp or screw band and hold the rim of the lid with your fingertips. If the seal is adequate, the lid will support the weight of the jar.

If a vacuum seal has not formed properly on any jar during processing, refrigerate the contents and eat within 1 week.

CASSOULET

Nick can make this in his sleep, and on occasion he has had to. There was a time, during our brief but glorious career as soup tycoons, when half the West End of London ate this cassoulet every lunchtime. Or that's how it seemed. We didn't used to can it, but we easily might have done, and with hindsight, we probably should have.

This recipe produces a tasty, warming and satisfying soup/stew which takes very well to life in a jar.

1.2 litres (2 pints) pork or chicken stock

600g (1¼lb) butter beans

1 duck breast

1kg (2¼lb) Toulouse sausage (or 'Taplows', see page 92)

1 carrot, diced

2 sticks of celery, diced

1 small onion, diced

40g (1½oz) pancetta, diced (see page 70)

2 tablespoons plain flour

200ml (7fl oz) tomato passata (see page 195)

4 bay leaves

A couple of sprigs of thyme

Salt and pepper

We make our stock from the spare ribs and skin of the pork belly we use for sausages. You could also use chicken stock if desired. The fat should be skimmed from the stock and reserved.

Soak the butter beans overnight, then simmer for 1½ hours or until soft.

Either hot smoke the duck breast for 1 hour at 110°C/230°F (see page 76) or pan-fry for 5 minutes on each side, reserving the oil. Dice the breast when cool.

Fry the sausages until lightly browned, then thickly slice when cool. Reserve all fat and juices.

Heat the pork stock in a small pan.

Take the reserved oils amassed during the preparation of this dish and pour 80ml (3fl oz) into a large saucepan. Fry the carrot, celery, onion and pancetta in the oil over medium heat for around 10 minutes or until soft.

Add the flour and stir in until it has soaked up all the fat. Then add the tomato passata, mixing it in well to remove any lumps. Slowly add the hot stock, stirring vigorously as you go, until it is all incorporated. Add the sausage, duck and beans, bay leaves and thyme. Simmer for half an hour, then season with salt and pepper to taste. This dish needs to be stirred frequently while cooking because the flour it contains is liable to stick.

PRESSURE CANNING

You can either hot- or raw-pack the cassoulet. We'd recommend hot. Leave a gap of 2.5cm (1in) between the surface of the stew and the lid of the jar. Process the cassoulet in your canner for 75 minutes if packed into ½litre/pint jars and 90 minutes if in litre/quart jars. The pressure settings should be as follows, depending on your altitude:

Sea level–710m (2000ft) – 0.76 bar/11 psi

710–1420m (2000–4000ft) – 0.83 bar/12 psi

1420–2130m (4000–6000ft) – 0.9 bar/13 psi

2130–3550m (6000–8000ft) – 0.97 bar/14 psi

Leave to cool at room temperature and store for up to 1 year.

BAKED BEANS

These are tangier than shop-bought baked beans and contain much less sugar, but they look pretty similar. It is a special thrill to make something so strongly associated with mass-produced aluminium tins in the home.

Once you have canned your beans, store them in a dark cupboard or larder. They will keep for a year.

THE BEANS

500g (1lb 2oz) dried haricot beans (known as navy beans in the USA)

THE SAUCE

3kg (6¾lb) ripe, sweet tomatoes, chopped

150ml (¼ pint) cider vinegar

10 cloves

4 green cardamom pods

½ teaspoon ground white pepper

½ teaspoon ground mace

½ teaspoon ground allspice

¼ teaspoon ground cinnamon

2 teaspoons paprika

150g (5oz) sugar

4 cloves of garlic, chopped

25g (1oz) sun-dried tomatoes (see page 28), finely chopped

Soak the beans overnight in cold water. Boil furiously for 10 minutes in slightly salted water, then simmer for a couple of hours until soft. Drain and reserve.

Simmer the sauce ingredients for 2–3 hours until reduced by at least 25 per cent. Pass the mixture through a food mill. This will yield about 1.6kg (3½lb) of sauce.

PRESSURE CANNING

Hot-pack the beans into jars, filling them three quarters full. Then pour over the hot sauce, leaving a headspace of 2.5cm (1in).

Pressure-can the beans for 65 minutes if they are in ½ litre/pint jars and 75 minutes if in litre/quart jars. The pressure settings should be as follows, depending on your altitude:

Sea level–710m (2000ft) – 0.76 bar/11 psi

710–1420m (2000–4000ft) – 0.83 bar/12 psi

1420–2130m (4000–6000ft) – 0.9 bar/13 psi

2130–3550m (6000–8000ft) – 0.97 bar/14 psi

TOMATO PASSATA

Passata is sieved tomato, in other words tomato minus the pips. Making it is one of the best ways to preserve your tomato crop. Because of its purity, passata can be used as the base for a myriad of sauces, for example bolognese. By preparing it yourself, you can reduce it until the flavour reaches the level of intensity you require.

As always, the end-product is only as good as its ingredients. Cherry tomatoes convert into excellent passata. Recently we made an unusually coloured version with yellow ones.

Take as many tomatoes as you like. Cut them in half and boil them in a large saucepan. Aim to reduce them by about 25 per cent. Pass the tomatoes through a food mill, making sure that the setting is small enough to eliminate the pips. Fill into jars leaving a 1.25cm (½in) headspace.

CANNING

You can use either canning method to process your passata. If you choose the boiling-water method, you may want to add lemon juice or citric acid at the rate of 2 tablespoons or ½ teaspoon respectively per 900ml (32fl oz). Tomatoes are borderline candidates for this type of canning and increasing their acidity renders them safe. It does, however, influence their taste. For this reason, we'd advise you to circumvent the problem by using a pressure canner, which will anyway yield a superior product as the process takes less time. But you may not have access to one...

In both cases, the jars should be hot-packed leaving headspaces of 1.25cm (½in).

For the correct processing times for each method, see the illustrative example on page 191.

GAZPACHO USING TOMATO PASSATA

This extremely healthy soup is essentially a liquid salad, given a zesty kick by the lemon juice and chillies. Our tomato passata is a vital ingredient, but if you haven't made your own, there are some reasonably tasty varieties on the supermarket shelves. Serves 4

1 litre (1¾ pints) tomato passata (see left)
2 teaspoons sugar
100ml (3½fl oz) balsamic vinegar
100ml (3½fl oz) olive oil
500ml (18fl oz) vegetable stock
3 slices of stale white bread
½ cucumber, roughly chopped
1 bunch of basil
1 bunch of flat leaf parsley
1 bunch of mint
Juice and zest of 1 lemon
2 red chillies, seeds removed and chopped
2 cloves garlic, chopped
1 teaspoon paprika
Salt
Freshly ground black pepper

- Blend all the ingredients together until thick and emulsified, either in a large bowl with a hand blender in or in batches in a food-processor.
- Chill for a couple of hours before serving.
- Serve with ciabatta bread and a mozzarella side salad.

LUNCHEON MEAT

Nick recently visited his in-laws in Malaysia. He was fed delicacies round the clock and primed to expect something really special for the farewell dinner. When his mother-in-law Honey disappeared into the kitchen, the rest of the family, knowing what was coming, started to quiver with anticipation. She returned with her *pièce de résistance*: luncheon meat fritters with corn. The assembled diners set upon it with relish.

At first, Nick found it hard to share in their enthusiasm. Back in his own country, luncheon meat is low-status stuff. Monty Python were ridiculing Spam for its naffness more than a generation ago. But when Nick tried the fritters, he found them spicy, delicious and utterly at home in the oriental setting. The primary taste instantly reconnected him with his childhood.

It turns out that luncheon meat is canned in countries as diverse as Korea, Denmark, Japan and the Philippines. Its popularity reflects the historical importance of spiced meat. Though this has now been reduced by improved transportation and refrigeration, old habits die hard.

Many people will have been wrongly but understandably put off luncheon meat by nasty 'economy' versions. Commercial varieties are often made with substandard or reclaimed meat, but there is no danger of that here. Nick has made the rehabilitation of luncheon meat a personal crusade.

TO MAKE 3 X 500G (1LB 2OZ) JARS
   1kg (2¼lb) pork shoulder mince
   1 medium onion, finely diced
   2 eggs, lightly beaten
   2 tablespoons plain flour
   2 teaspoons salt
   2 cloves of garlic, chopped
   ½ teaspoon white pepper
   1 teaspoon mustard powder
   ½ teaspoon ground allspice
   ¼ teaspoon ground cloves
   ½ teaspoon ground mace
   ½ teaspoon ground ginger
   1 tablespoon Worcestershire sauce
   500ml (18fl oz) chicken or pork stock

Mix all the ingredients together in a large bowl. Empty into a rectangular ovenproof dish, then cover and bake for 2 hours at 160°C/320°F/gas mark 3.

Remove from the oven and before the meat is cool, cut it into sections and fill into jars. Top up with hot stock, leaving a 2.5cm (1in) headspace.

PRESSURE CANNING
Pressure-can the luncheon meat for 75 minutes if packed into ½ litre/pint jars and 90 minutes if in litre/quart jars. The pressure settings should be as follows, depending on your altitude:

Sea level–710m (2000ft) – 0.76 bar/11 psi
710–1420m (2000–4000ft) – 0.83 bar/12 psi
1420–2130m (4000–6000ft) – 0.9 bar/13 psi
2130–3550m (6000–8000ft) – 0.97 bar/14 psi

This luncheon meat is also nice straight out of the oven. It is very good in stir-fries or fried and served in a sandwich with roasted peppers.

LUNCHEON MEAT AND CORN FRITTERS

When our friend Dominic heard about the central ingredient of this recipe, he was concerned. 'But that's fishing bait mate!', he said to Nick. This only redoubled the latter's resolve. Here is the testimony of a paid guinea-pig (his wife): 'These fritters possess an explosive powdery crunch and a soft, light centre. Immediately after the first bite, one's tastebuds are caressed with a deep hamminess, tempered only by the juicy sweetness of the sweetcorn. The aromatic aftertaste of coriander has left a wonderful feeling in my mouth.' Serves 4

1 tablespoon corn flour
75g (3oz) plain flour
½ teaspoon baking powder
oil, for deep frying
200g (7oz) tinned sweet-corn, drained (reserve the drained juice)
250g (9oz) luncheon meat, cut into small cubes

1 tablespoon fish sauce
2 tablespoons juice from the bottom of the can of sweetcorn
1 tablespoon chopped coriander
2 spring onions, chopped

- Mix the corn flour with the plain flour and baking powder in a large bowl. Add the rest of the ingredients and combine thoroughly with a wooden spoon.
- Heat the oil to frying temperature (180°C/356°F) and fry the fritters in small batches, portioning them with a tablespoon and slipping them into the hot oil. Fry on each side for around 3–4 minutes until golden brown.
- Serve with sweet chilli sauce (see page 177).

TONGUE

Once you get over the strangeness of having an animal's tongue in your mouth, you may find that you like it quite a lot. The flavoursome flesh has a unique consistency, dense yet soft and slightly gelatinous. Ironically, it melts in the mouth.

A whole beef tongue is quite a spectacular item. To successfully process and preserve one gives a considerable sense of achievement.

1 beef tongue
1.5 litres (2½ pints) beef stock
4 bay leaves
2 sprigs of fresh thyme
6 cloves
Salt and pepper
½ teaspoon ground allspice
1 onion, cut in half
1 carrot, peeled
2 sticks of celery

Wash the tongue, then simmer with the other ingredients for 2½ hours until tender. While still hot, cut into chunks and stuff into jars. Bring the stock to the boil, remove the solid ingredients, then pour over the meat, leaving a 2.5cm (1in) headspace.

PRESSURE CANNING

Process the tongue for 75 minutes if packed into ½ litre/pint jars and 90 minutes if in litre/quart jars. The pressure settings should be as follows, depending on your altitude:

Sea level–710m (2000ft) – 0.76 bar/11 psi
710–1420m (2000–4000ft) – 0.83 bar/12 psi
1420–2130m (4000–6000ft) – 0.9 bar/13 psi
2130–3550m (6000–8000ft) – 0.97 bar/14 psi

PEACHES

Bottling does great things for peaches. They become gloriously slippery and glossy, and they infuse the syrup with their flavour. Bottled peaches scream out for ice cream. To make a simple Peach Melba, combine them with raspberry ripple.

> **350g (12oz) sugar**
> **2kg (4½lb) peaches**
> **1 litre (1⅔ pints) water**

Heat the sugar in a pan with the water until fully dissolved.

Prepare a pan of boiling water and a bowl of iced water. Dip the peaches into the boiling water for 30 seconds, then plunge them into the cold. This should loosen the skins. Peel the fruit, then halve and remove the stones.

Pack the peaches quite tightly into the jars to just below the neck, then add the hot syrup so that they are completely covered. You should leave a headspace of 1.25cm (½in).

BOILING-WATER CANNING

This time you will be using the simpler 'boiling-water' method of canning. Once again, processing time depends on your altitude and the size of the jars.

> 0–305m (0–1000ft) – boil ½ litre/pint jars for
>   20 minutes, litre/quart ones for 25
> 305–915m (1000–3000ft) – boil ½ litre/pint jars for
>   25 minutes, litre/quart ones for 30
> 915–1625m (3000–5000ft) – boil ½ litre/pint jars for
>   30 minutes, litre/quart ones for 35
> and so on...

Allow the peaches to cool at room temperature, then store in a cool cupboard or larder for up to 1 year.

PEACHES SWIMMING IN BLACKBERRY SAUCE

This can be made with bottled peaches, peaches preserved in alcohol or indeed the fresh fruit. Serves 4

8 peach halves, bottled, preserved in alcohol (see left) or fresh

THE BLACKBERRY SAUCE
300g (10½ oz) frozen blackberries
50ml (2fl oz) port
125g (4½oz) sugar
Juice of ½ a lemon

THE SPONGE
75g (3oz) unsalted butter, softened
75g (3oz) caster sugar
1 medium egg, plus 1 egg yolk
50g (2oz) flour
1 teaspoon baking powder
50g (2oz) ground almonds
100ml (3½fl oz) double cream
150g (5½oz) peach jam

- To make the blackberry sauce, simmer the blackberries with the port, sugar and the lemon juice for 10 minutes, then attack with a potato masher. Pass through a conical strainer, discard the seeds and reserve.
- To make the sponge, cream the butter and the caster sugar together with a whisk until white and light.
- Lightly beat the egg and yolk together. Add a small amount to the butter and caster sugar and whisk in. Continue until all the egg is gone; if you go too fast, the mixture will curdle.
- Sift the flour and baking powder into a bowl containing the ground almonds. Stir thoroughly, then whisk a little of this dry mixture into the egg/butter/sugar sponge mix.
- Now whisk in a little of the cream, then more flour mix and so on. When both are incorporated, fold in the jam.
- Grease 10 shallow bun tins with a little butter or line them with paper cases. Divide the sponge evenly between them and bake in the oven for around 20 minutes at 180–190°C/350–375°F/gas mark 4–5.
- Pour a pool of blackberry sauce onto each plate, then top off with the peaches and sponge. Serve with the cream.

AIR EXCLUSION

Successful food preservation has two elements. The first is the killing or dramatic suppression of potentially harmful organisms present in the food itself. This is the primary purpose of all the drying, salting, smoking, pickling, boiling and fermenting we have encountered in previous chapters. But the next line of defence is to prevent dubious microbes getting at the food in the first place – or the second if something has been done to it to kill off the nasties in round one. This applies to all the items in this chapter, even if all that has happened to them is a period of cooking.

The simplest and most ancient technique of putting up a barrier against airborne organisms is to pack the relevant food in some (preferably edible) substance that they are unable to get their teeth into. Not so much, as it turns out, because they find the fats typically employed inedible, but because most of them are killed during the cooking phase, and the rest are unable to get in, or to move and hence breed, once the fat has congealed. This is the rationale behind the layer of butter on top of a delicate pâté, or the creamy duck fat in which *confit de canard* is entombed. The most beautiful dishes preserved in this way are those relying on clear mediums such as aspic, which hold them in highly visible suspended animation.

One of the oldest preserved foods relying on the principle of air exclusion is Lebanese *qawrama*. This is a kind of *confit* made from the browned, lean meat of the fat-tailed sheep. This is combined with rendered fat from the animal's abundant tail and sealed in special preserving jars. Fat-tailed sheep were already well established in Arabia by the middle of the first millennium BCE. Amazed Greek authors reported that the appendages that gave them their name were sometimes so large that they had to be supported by little carts.

In Britain, potted meats became all the rage during the late seventeenth century, once the technology was in place to mass-produce suitable vessels. All kinds of seafood and game were preserved in this manner, as were domestic products like pork and beef. The food was liberally spiced with pepper, nutmeg and cloves, and packed into earthenware tubs sealed with clarified butter or pork or duck fat. These potted products were the forefathers of the soft pastes that today fill the sandwiches of a million small children. Some would keep for several months.

The other way, of course, of preventing air-borne microbes getting at your food is to surround it with literally nothing. Vacuum packing may lack the glamour of the venerable methods covered elsewhere in this book, but for keeping preserved items in pristine condition, it is unrivalled.

VACUUM PACKING

If food is encased in a plastic wrapper moulded to fit its contours, airborne organisms will be unable to get in and aerobic organisms inside will be denied the oxygen they need to function. Vacuum packing is much used in the catering industry for preserving purposes. In essence, it works like this: the food is placed between two sheets of plastic and these are sealed together on three sides with hot pressing irons. This makes a 'bag', out of which the air is then sucked with a pump. Then the fourth side of the 'bag' is sealed before air has a chance to get back in.

Vacuum packing is eminently feasible in the home and it will greatly enhance your preserving options. Food packaged in this way will keep in the fridge for at least twice as long as usual. If it is frozen, the preserving effect is even more pronounced. Fish will keep for up to 2 years instead of the usual 2–3 months, and red meats for as long as 5 years. Oily foods will not become rancid or bitter tasting.

A new home vacuum packer will set you back a good £100, but you may be able to find a decent second-hand one. The process is fun to watch. When the air is evacuated, the plastic suddenly clings to the food like a second skin.

ASPIC

As 'preserved in aspic' is a common expression in everyday English, it would be foolish of us not to say something about it here. But we should point out that aspic on its own only preserves food very briefly (items tinned in it are another matter). The familiar metaphor refers more to the beauty of food preserved in aspic than to its longevity.

Aspic in its purest form is the clear, protein-rich jelly that forms under a cold roast chicken or a refrigerated ham. In practice, it is usually made from meat or fish stock. This is often reduced to ensure a better set, or gelatin is added to it. The outcome is a delicate savoury jelly. Aspic makes an attractive glaze, helping to prevent the food it coats from drying out, and in larger quantities it can hold morsels in apparent defiance of gravity.

Johnny's great-aunt, who brought him up, made the following starter every time she cooked a big meal. It looked as though it must have taken ages to make, but in fact was virtually instant. It went down so well that she never felt the need to deviate.

EGGS AND PRAWNS IN ASPIC (FOR EACH DINER)
A small handful of cooked peeled prawns
A couple of chives, chopped
200ml (7fl oz) good tinned beef consommé
A dash of sherry
½ hard-boiled egg
1 ramekin

Mix the prawns and chives with the consommé and sherry.

Fill the bottom third of each ramekin with the mixture and gently place the egg in it, curved side down. Then spoon over the remaining consommé so that the egg is well covered. Smooth the surface with a spatula.

Chill and serve. For added effect, garnish each ramekin with a prawn and a few lengths of chive.

CONFIT DE CANARD

This method of preserving duck legs and thighs in fat originates from south-west France. The meat is soft, dark and intensely flavoured. You may end up licking your plate.

Salt
6 duck legs
Cracked black pepper
3 cloves
2 cloves of garlic
750g (1lb 10oz) goose or duck fat, lard, olive oil, or a combination of all 4

Sprinkle salt onto the exposed flesh of each duck leg. Fold the legs together so the salted areas are touching and leave them in the fridge for 24 hours.

Wipe the legs down and firmly place them in an ovenproof dish with a tight-fitting lid, together with the pepper, cloves and garlic.

Cook in the oven at 180°C/350°F/gas mark 4 for 2½ hours.

Leave the legs to cool in their own fat, then pack them tightly into sterilised jars (see page 164). Heat the additional fat/oil/lard (i.e not that released by the duck) and when molten, pour it over the duck up to the necks of the jars. Seal (see page 164), store in the fridge and eat within a fortnight.

GOOSE RILLETTES

The French are less uptight about fat than most nationalities. They recognise its role in a healthy diet and are careful to distinguish between good and bad kinds. Above all, they know that enjoying your food can put years on your life, offsetting any theoretical health risks. Jeanne Calment, the longest-lived human being of all time, didn't get to 122 on a diet of beansprouts.

Rillettes consist of seasoned meat slowly cooked in fat, then teased apart and preserved in it.

1 goose, weighing around 4kg (9lb)
800g (1¾lb) pork shoulder, boned
600g (1¼lb) pork belly fat
3 bay leaves
3 thyme branches
8 black peppercorns
Salt
6 juniper berries

Remove the skin from the goose. De-bone the bird, then cut the meat into chunks, reserving any fat you come across.

Mince together the pork shoulder, belly fat and goose fat.

Combine all the ingredients in a large, heavy-based pan. Cook slowly for 3 hours with the lid on, stirring occasionally. Don't let the mixture stick. If it starts to, add a drop or two of water.

The goose meat will eventually start to fall apart. You can hasten the process by teasing it with a fork.

Sterilise some pots (see page 164). Take out the goose meat with a slotted spoon and press it down firmly into them. Pour the remaining fat on top, cover and leave to set in the fridge.

Serve with warm French bread and cheap rosé wine. The rillettes will keep for at least 2 weeks in the fridge.

CHICKEN LIVER PATE WITH MORELS

This is as much about texture as taste, with nuggets of morel interrupting the luxurious smoothness.

- 10 medium-sized fresh morels, or any other good edible mushrooms
- 200g (7oz) unsalted butter
- 400g (14oz) chicken livers (or duck if preferred)
- 2 cloves of garlic, finely chopped
- 50ml (2fl oz) brandy
- 1 teaspoon mustard powder
- Salt and finely ground white pepper
- 1 sprig of thyme, finely chopped
- ½ teaspoon freshly grated nutmeg
- 50ml (2fl oz) double cream
- 200g (7oz) clarified butter (see 'Potted Shrimps' on page 206)

Roughly dice the mushrooms and gently fry them in 30g (1oz) of the unsalted butter for 5 minutes.

Fry the liver and garlic in 50g (2oz) unsalted butter over medium heat until cooked through. This will take 8–10 minutes.

Take the livers out with a slotted spoon and place them in a blender. De-glaze the pan with the brandy and add this liquid to the livers.

Melt the rest of the unsalted butter in the pan and pour it into the blender. Add the mustard powder, salt, pepper, thyme, nutmeg and double cream.

Blend until smooth. Then pour into a mixing bowl. Throw in the mushrooms and mix with a spoon until evenly distributed.

Spoon the mixture into a sterilised earthenware dish (see page 164) and smooth the surface with a palette knife. Leave the pâté in the fridge for an hour or so to set.

Heat the clarified butter and pour it over the pâté so that the surface is covered.

Cover and return to the fridge. This pâté will have a shelf-life of 7 days in the fridge before the seal is broken. After breaking the layer of clarified butter, consume within 3 days.

WILD MUSHROOM PATE

We'll finish with this nice vegetarian pâté, a prudent thing to make after a successful afternoon's foraging.

- 500g (1lb 2oz) chanterelle or oyster mushrooms, sliced
- 60g (2½oz) butter
- 2 medium shallots, chopped
- 1–2 cloves of garlic, crushed
- 25g (1oz) dried porcini mushrooms (see page 21), cleaned but unsoaked
- 100g (3½oz) dried green lentils
- 100g (3½oz) cream cheese
- 100 ml (3½fl oz) crème fraîche
- Salt
- Cracked black pepper
- 1 sprig of flat-leaf parsley, chopped
- 200g (7oz) clarified butter (see 'Potted Shrimps' on page 206)

Fry the fresh mushrooms in the butter with the shallots and garlic. They will release lots of juice. Gently cook until this has reduced by at least half.

Add the dried porcini to the lentils and just cover with water. Gently cook until all the water has been soaked up. Check the lentils to see if they are ready. If not, add a little more water and continue cooking until they are.

Add the lentils to the mushrooms and blend in a food-processor until smooth.

Add the cream cheese, crème fraîche, salt, pepper and chopped parsley.

Spoon the pâté into a suitable, sterilised container (see page 164). Warm the clarified butter and pour it over. Cover and leave to set in the fridge.

The pâté will keep for 5 days in the fridge. After the butter seal has been broken, consume within 3 days.

POTTED SHRIMP

After a bereavement, a kindly neighbour asked Johnny if she could do anything for him. 'You wouldn't mind picking up some potted shrimps for me in town?' he replied.

Potted shrimps are very comforting. They are delicate, mildly spiced and redolent of a bygone era. The crustaceans in question are the small, grey-to-brown kind caught with a shrimping net on old-style British seaside holidays. You can buy them ready prepared, but this is expensive.

100g (3½oz) unsalted butter
¼ teaspoon cayenne pepper
2–3 blades of mace
A good grating of nutmeg
400g (14oz) cooked, peeled brown shrimps
150g (5oz) clarified butter*

*To clarify the butter, slowly melt it in a pan. Skim the scum from the surface, then ladle off the clear, clarified butter, leaving behind the watery residue which will have collected at the bottom of the pan. Pour the butter into a jar and store in the fridge.

Heat the unsalted butter in a pan until melted but not boiling. Add the cayenne pepper, mace and nutmeg and stir in.

Pour in the shrimps and gently cook for 5 minutes to re-sterilise them.

Fill the shrimps and spiced butter into 2 sterilised pots or ramekins (see page 164) of around 300ml (½ pint) capacity. Press them down to compact them.

Place in the fridge until set. This will take about 30 minutes.

Heat the clarified butter and pour it onto the shrimps. Once set, it will form an impermeable seal.

The shrimps will have a shelf-life of 2 weeks in the fridge so long as the butter seal remains intact and 3 days once broken.

Serve warmed with toast and squeeze on lemon juice to taste.

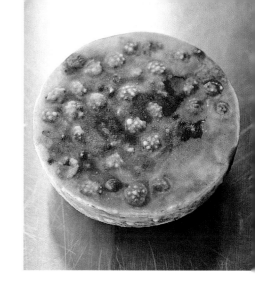

FREEZING

Secreted in the woodland near where Johnny lives is what looks like a wildly overgrown stone igloo. Today it is only used by children playing hide-and-seek or teenagers up to something less innocent. But it is, in fact, a Victorian ice-house. Long before the invention of the refrigerator, wealthy Europeans and Americans were building these edifices and filling them with ice brought by ship from the far north to enable them to chill food during the summer. Often their primary purpose was to meet the already insatiable demand for ice-cream. But the Victorians were not the first to build ice-houses. The Mesopotamians were at it four thousand years ago, and the Egyptians, Greeks, Persians and Chinese, none of whose summers was exactly cold, were not far behind. The Inuit, meanwhile, have been aware of the preservative powers of ice since before time began. Indeed, some families in the Arctic have been known to use refrigerators to *prevent* their food from freezing, as they keep it a couple of degrees above zero (or 32°F). Similarly, residents of the altoplano in the Andes have been freeze-drying potatoes for eons. They crush their spuds and spread them on rocks at altitudes in excess of 15,000 feet. At night, when the temperature falls well below zero, they freeze into *chuno* which is then made into flour. An alternative method involves dipping whole potatoes in water, leaving them out overnight and then trampling the moisture out of them with bare feet (!).

Such techniques were all very well for residents of the Arctic or Andes, or for those in warmer climes rich enough to import ice from distant snowy regions. But the race was on to find a practical solution for everyone else. One promising idea was to harness the power of evaporation to conduct heat away from a receptacle of liquid. The Ancient Egyptians had made many a slave stay up all night wetting the outside of earthenware jars to cool the water within. But this approach was hard work.

The eventual answer, after a phase dominated by complicated devices relying on repeatedly condensing then evaporating gases like ammonia, depended on the skilful use of electricity. In 1925 Clarence Birdseye unveiled his patented 'Quick Freeze Machine'. While working as a fur trapper in icy Labrador, he had noticed that food frozen at –40°C and below tasted remarkably fresh when thawed. This was because the process happened so quickly that large, damaging ice crystals were unable to form. After years of working out how to duplicate such temperatures in the warmer south, he had paved the way for everyone to have an Arctic in their home.

Home freezing may seem straightforward, but there are a couple of important things to bear in mind. Firstly, the process doesn't actually kill micro-organisms, it just renders them dormant. Once the food is defrosted, they spring back into life. Secondly, great care needs to be taken with the preparation of foods for the freezer. The process must be rapid to minimise tissue damage. Full instructions are given below.

BLANCHING

Living plants contain enzymes which orchestrate their growth and maturation. When a vegetable is picked, its enzymes are still active, and unless something is done about them, they will continue to influence it. This can have adverse effects on taste, texture and appearance. Asparagus, for example, will start to taste grassy. Its stalks will toughen and their colour will fade.

Briefly immersing vegetables in boiling water or steam halts the action of their enzymes, and this is the rationale behind blanching them prior to freezing. Almost all vegetables benefit from the procedure. Here are some basic tips:

Use as large a pot of boiling water as possible – at least 5 and preferably 10 litres (1 or 2 gallons). This will minimise the time the water comes off the boil when you add the vegetables. This helps them to retain their colour.

Fill a very clean sink with iced water. Once vegetables have been blanched for the prescribed period they need to be chilled immediately or they become faded, soft and pulpy.

The quicker you freeze vegetables, the better. If you plan to process a large batch, turn down your freezer temperature in advance to prepare it for the extra load. Make sure you don't overload or the vegetables will take an inordinate time to freeze.

CUTTING AND PORTIONING

Whatever you are freezing, you need to think hard about how you are going to package it. The problem is that once you have frozen it, it will be very hard. As hard as ice, in fact. Unless you do a little forward planning, you will end up with a gigantic, unwieldy block of vegetables welded together with ice.

The solution is to initially freeze or part-freeze the vegetables in small, individual clusters. To do this, use the tray method:

Take the largest metal or plastic tray that will fit in your freezer and arrange the contents to accommodate it. It needs to sit flat and stable and to have a little headspace.

Blanch the vegetables if appropriate, then dry them thoroughly on kitchen paper. If you leave their surfaces wet, they will stick to their surroundings as they freeze.

Now arrange the vegetables into small mounds on the tray. These should weigh about 50g (2oz) each.

Carefully transport the tray to the freezer, making sure that the piles remain intact and separate, and leave inside until frozen.

Remove the tray and transfer the individual clumps of vegetables into freezer bags. A fish or cake slice will help you separate the mounds from the trays. If they are stuck fast, run some cool water over the other side of the tray to help dislodge the portions.

Label the bags and store in the freezer.

Some vegetables need to be cut (e.g. courgettes) or broken up into florets (e.g. broccoli) before you freeze them. You won't be able to do it afterwards. Recommendations for specific items are given on page 215.

USEFUL EQUIPMENT

A very large pan

Large metal or plastic trays

A long-handled 'bird's nest' or large slotted spoon (for removing vegetables from boiling water)

Ice

Labels

Kitchen paper

Resealable freezer bags

Plastic containers with tight-fitting lids (these should be rectangular rather than round for efficient storage)

DUXELLE OF MUSHROOMS

This is a crafty way of preparing mushrooms for the freezer. A 'duxelle' is simply a finely chopped vegetable (let's not get into the taxonomic status of fungi) which in this case is cooked up with garlic, shallots and herbs. The technique is effective with many different kinds of mushroom; you can experiment with different mixes as well as making 'single species' duxelles, for instance with chanterelles.

The other day Nick added a couple of frozen blocks of porcini/cep/boletus duxelle to a carbonara sauce. The outcome was, needless to say, delicious.

> 50g (2oz) unsalted butter
> 2 shallots, peeled and finely chopped
> 1 or 2 cloves of garlic, finely chopped
> 500g (1lb 2oz) mixed mushrooms, finely chopped
> Salt and pepper
> 50ml (2fl oz) chicken or vegetable stock
> A sprig of parsley, chopped
> A few more herbs if desired, such as thyme, chives
>   or dill, chopped

Melt the butter in a pan and gently fry the shallots and garlic until softened.

Add the mushrooms and a little salt and pepper. Fry until the 'shrooms have released their juices.

Add the stock and reduce until there is very little left at the bottom of the pan. Finally, stir in the herbs.

Store in the fridge if you're planning to use within 24 hours, otherwise pour into ice trays and freeze. The duxelle will keep for several months once it is frozen.

CEP AND SPINACH SALAD WITH GINGER VINAIGRETTE

Ceps (*Boletus edulis*) are the mushroom-hunter's holy grail. Their dense, meaty flesh stands up particularly well to the freezer. Small specimens can be frozen whole. They should be cooked directly from the freezer, although you'll need to let them partially defrost if you plan to slice them.

Any frozen or for that matter fresh mushroom could be substituted for the ceps. **Serves 2**

THE DRESSING

2 teaspoons honey

2 teaspoon sesame seeds, dry-fried for 2 minutes until slightly toasted

Juice of 1 lime

1 tablespoon sesame oil

1 tablespoon tamari

2 tablespoons chopped ginger

THE SALAD

8 small frozen ceps

1 tablespoon vegetable oil

Small knob of butter

250g (9oz) baby leaf spinach, washed thoroughly

1 medium carrot, roughly grated or cut into a julienne

3 spring onions, finely sliced

Place all the ingredients for the dressing in a small container with a tight-fitting lid and shake vigorously.

Take the ceps out of the freezer. When they have slightly defrosted, cut them into quarters. Heat up the vegetable oil in a frying pan and fry the mushrooms over moderate to fierce heat until browned nicely on each side. Reserve.

Melt the butter in the same pan, then add the spinach and wilt it by cooking until it has halved in volume.

Mix the carrot julienne with the spinach in a large bowl, then pile the salad onto the plates and lay the cep quarters on top. Garnish with spring onions.

Shake the dressing again, add about 1 tablespoon to each plate, then serve.

FREEZING INSTRUCTIONS FOR INDIVIDUAL VEGETABLES

All vegetables must be washed thoroughly before freezing, then blanched. Either plunge into a large quantity of water on a rolling boil for the time specified, or steam them for 50 per cent longer.

ASPARAGUS
Only freeze tender asparagus. Trim off the stalk ends and cut into sections if desired. Blanch medium asparagus for 3 minutes, thick spears for slightly longer and thin ones for slightly shorter. Cool promptly, then freeze.

BEANS (GREEN AND RUNNER)
Trim and slice runner beans. Blanch either kind for 3 minutes.

BEETROOT
Boil for 40 minutes. Cool, then peel and slice before freezing.

BROCCOLI AND CAULIFLOWER
Separate into florets and blanch for 3 minutes before cooling. Tray freeze before packaging.

BRUSSEL SPROUTS
Trim and boil for 5 minutes before cooling and freezing.

CABBAGE AND CHINESE GREENS (E.G. PAK CHOI AND CHOI SUM)
Slice and blanch for 2 minutes before cooling and freezing.

CARROTS
Blanch them for 5 minutes if small and whole or 2 minutes if sliced or cut into strips.

CORN ON THE COB
Blanch for 9 minutes if you want to freeze them whole. Otherwise, blanch for 3 minutes, chill, then shave the kernels from the cobs. Bag them and freeze them.

COURGETTES
Blanch them whole for 4 minutes, adjusting this time if they're particularly thin or thick. Slice when cool, then freeze.

GREENS (INCLUDING SPINACH AND CHARD)
Blanch for 2 minutes, then chill and drain well. You can remove most of the moisture by putting the greens in a sieve and pressing against them with a spoon.

MUSHROOMS
There are various ways of freezing mushrooms, depending on the species. See recipes for Duxelle of Mushrooms (page 213) and Cep and Ginger Salad (page 214).

ONIONS
Do not need to be blanched. You can just chop them and freeze them.

PEAS
Blanch for 1½ minutes before freezing.

PEPPERS
Again, you don't need to blanch them. Just de-seed and chop before freezing.

NB: All vegetables should be cooked straight from the freezer.

FREEZING FRUIT

There is much to be said for using the tray method (see page 212) when freezing fruit. This will make it easier to remove the desired quantity from the bag when you wish to use them.

Freezing can harm the structure of some fruits. Although this won't necessarily impair their flavour, the damage can be minimised by carrying out the freezing as quickly as possible after picking.

SUGARING

Some fruits survive better in the freezer if they are lightly sugared beforehand. This is true of blackberries, strawberries, blackcurrants, redcurrants and plums. Other soft fruits, including blueberries and raspberries, freeze perfectly well as they are. If you are sugaring fruit prior to freezing, aim for a light, even coating.

In practice, Nick often cooks his fruit before freezing. With plums, for example, he will remove the stones, add a little sugar, cook until soft, then batch up and freeze.

Fruit purées and coulis can also be frozen. To make a coulis, boil the fruit with 25 per cent of its volume of caster sugar, then sieve.

RECIPES

Freezing is not only useful for preserving food. It is also an integral part of the preparation of some of our favourite desserts and snacks. Where would we be without ice-cream? During the Second World War, the US government deemed it an 'essential foodstuff' crucial to military morale. American air crews sometimes made it by hitching large cans filled with ice-cream mix to their aircraft. During a sortie, the constant motion and extreme cold at altitude would do the work for them.

We're not going into the mysteries of ice-cream making here, but we do provide you with recipes for two simple frozen treats. One is a dinner party item, the second more of a snack.

ELDERFLOWER SORBET WITH CHAMPAGNE

If you've ever walked through an elderflower wood in spring, you'll know that the flowers have a pretty heady aroma. You don't need many to flavour whatever concoction you're working on. A few heads are enough for this powerful yet delicate sorbet.

250g (9oz) caster sugar
250ml (9fl oz) water
10 flowering elderflower heads
300ml (11fl oz) champagne
Juice and zest of 1 lime
2 egg whites

Combine the sugar and the water in a medium-sized saucepan. Boil for 5 minutes, then take off the heat and add the elderflowers. Leave to infuse for 1 hour, then strain off the liquid and discard the elderflowers.

Add the champagne, lime juice and zest, then freeze in a plastic container, stirring occasionally

When the sorbet is frozen but not completely hard, take it out of the freezer and cut it into rough chunks. Then blend them with the egg whites. This will take about 1 minute in a food-processor.

Pour the mixture back into the container and immediately restore to the freezer.

Take the sorbet out of the freezer 10 minutes before you mean to eat it.

Try serving in shot glasses as an elegant palate cleanser.

RIGHTEOUS RASPBERRY LOLLIES

The genius of these lollies is that they contain no dairy products or refined sugar. You can suck on them with a clear conscience and allow your children to do the same. And neither should have cause for complaint. Nick made some of these lollies from a collection of wild and albino raspberries assembled in the Pyrenees and on the Wicklow Way.

300g (10½oz) raspberries
200g (7oz) clear honey
Juice of 1 medium lemon
Lolly moulds (available in catering shops and
   department stores)
ice-cream sticks*

*You can buy these in catering shops and department stores. Alternatively, use wooden kebab skewers cut to size.

Heat the raspberries in a saucepan with the honey and lemon juice until the mixture comes to the boil. Remove from the heat and mash with a potato masher.

Pass the mixture through a vegetable mill or conical sieve to get rid of the pips.

Leave it to cool, then pour into the lolly moulds. Insert the lolly sticks, then freeze.

USEFUL ADDRESSES

The following are useful sources of advice, equipment and other supplies:

DRYING
In the UK:
www.ukjuicers.com
Sells a small selection of food dehydrators online

In the US:
www.drystore.com
US-based company selling dehydrators, vacuum-packers etc

SMOKING
We showed you how to build both a cold- and a hot-smoker in chapter 3, but you may prefer to buy one:

In the UK:
www.coldsmoker.com
makers of the 'West Country Cold-smoker'. Ship throughout the EC.

Gerry's of Wimbledon
170 The Broadway, London SW19 1RX
Tel: (020) 8542 7792
Suppliers of Shakespeare and Abu hot-smokers. Mail order.

In Canada and the USA:
Wells Can Company Limited
8705 Government Street, Burnaby, BC,
V3N 4G9 Canada
Tel: (604) 420 0959
www.wellscan.ca
Online vendors of the Bradley Smoker, a nifty machine capable of both cold- and hot-smoking.

SAUSAGES
In the UK:
The Natural Casing Company
High Point, Dippenhall, Farnham,
Surrey GU10 5EB
Tel: 01252 850 454
This company also sells saltpetre.

In the US:
www.butcher-packer.com
Suppliers of casings, starter cultures, curing mixes and sausage-stuffing machines. Also smokers, thermometers and vacuum-packing equipment.

CANNING
Home Canning Supply
PO Box 1158-WW, Ramana,
CA 92065, USA
Tel: (760) 7880520
www.homecanningsupply.com
US-based company, but ships internationally. Sells canners, dehydrators, jars etc.

www.canningpantry.com
www.polsteins.com
www.homesteadharvest.com
These three companies all sell canning equipment but do not currently ship outside North America.

SUGAR – JAM-MAKING EQUIPMENT
In the UK:
Wares of Knutsford
36a Princess Street, Knutsford, Cheshire
www.waresofknutsford.co.uk
On-line vendors of jam jars, funnels, straining bags etc.

In the US:
www.kitchenkrafts.com
Sell a wide range of jam-making, home-canning and other preserving equipment.

FERMENTING
Koji starter culture for making miso (page 148) is available online and shipped internationally from:

GEM Cultures
30301 Sherwood Road, Fort Bragg,
CA 95437, USA
Tel: (707) 9642922
www.gemcultures.com

AIR EXCLUSION – VACUUM SEALERS
www.kitchenkrafts.com
They will ship worldwide.

In the US:
www.butcher-packer.com
www.drystore.com
www.homesteadharvest.com

MISCELLANEOUS
The UK-based firm Wild Harvest are friends of ours who supply game and other high-quality meats as well as first-rate wild mushrooms. They have an efficient home-delivery service.

Wild Harvest Ltd
Units B61–4 New Covent Garden
Market, London SW8 5HH
Tel: (020) 7498 5397
www.wildharvestuk.com

INDEX

To Amy Mei

Many people have helped or inspired us in the writing of this book. In particular, we'd like to thank: Hugh Fearnley-Whittingstall for his generous foreword, Janie Suthering for her helpful suggestions and advice, The Women's Institute and the National Trust for pickling tips, Simon Benning, James Harder, Valentino and Maria Rosa for help with sausages, Justin Percival for directions to an inexhaustible supply of sloes and Hazel Holt for her microbial knowledge. We must also thank Keiko (sugar), Nigel (figs), Camilla (plums), Jane (paté), David (biltong) and Penelope (quinces) for help with specific items. Muna Reyal has been an endlessly patient and good-humoured editor, Carl Hodson has done a great design job and Peter Cassidy has produced some truly wonderful photography, aided by Linda Tubby's brilliant interpretations of the recipes.

First published in Great Britain 2004 by
Kyle Cathie Limited
122 Arlington Road, London NW1 7HP
general.enquiries@kyle-cathie.com
www.kylecathie.com

10 9 8 7 6 5 4 3 2 1

ISBN 1 85626 532 3

Text © 2004 Nick Sandler and Johnny Acton
Photography © 2004 Peter Cassidy
Book design © 2004 Kyle Cathie Limited

Senior Editor Muna Reyal
Designer Carl Hodson
Photographer Peter Cassidy
Food stylist for the recipe photography Linda Tubby
Styling Wei Tang
Editorial assistant Jennifer Wheatley
Production Sha Huxtable and Alice Holloway

A Cataloguing In Publication record for this title is available from the British Library.

Colour reproduction by Sang Choy
Printed and bound in Singapore by Tien-Wah Press